Forecasting
Urban Water Demand

Forecasting Urban Water Demand

R. Bruce Billings
Department of Economics
University of Arizona
Tucson, Arizona

and

C. Vaughan Jones
Economic Data Resources
Boulder, Colorado

1996

American Water Works Association
Denver, Colorado

Dedicated to Patricia, Edessa, and Carol,
and to water resource analysts and students
of forecasting.

Editor: Bill Cobban
Cover design: Scott Nakauchi-Hawn
Book design and layout: Alan Livingston

Published by
American Water Works Association
6666 W. Quincy Ave.
Denver, CO 80235

Library of Congress Cataloging-in-Publication Information

Billings, R. Bruce.
Forecasting urban water demand / R. Bruce Billings and C. Vaughan Jones.
xv, 179 p. 17×25 cm.
Includes bibliographical references and index.
ISBN 0-89867-827-7
1. Municipal water supply–Management. 2. Municipal water supply--Forecasting--Statistical methods. I. Jones, Clive Vaughan.
II. Title.
HD4456.B55 1996
333.91'12–dc20 95-44740
CIP

Lotus 1-2-3® is a trademark of Lotus Development Corporation.
Excel® is a trademark of Microsoft Corporation.

Contents

Figures . . . xi
Tables . . . xiii
Acknowledgments . . . xv

Chapter 1: Introduction . . . 1
- Why Forecast? . . . 2
- Major Uses of Forecasts . . . 3
  - Long-Term Water Demand Forecasts . . . 3
  - Medium-Term Water Demand Forecasts . . . 4
  - Short-Term Water Demand Forecasts . . . 4
- Major Influences on Water Demand . . . 4
  - Population, Economic Cycles, and Technology . . . 4
  - Weather and Climate . . . 4
  - Price . . . 5
  - Efficiency and Conservation Programs . . . 5
  - Other Factors . . . 5
- Water Demand Forecasting Approaches and Techniques . . . 6
  - Per Capita and Other Unit Use Coefficient Approaches . . . 6
  - End-Use Models . . . 6
  - Extrapolation . . . 7
  - Causal/Structural Models . . . 7
  - Choice of Methods . . . 8
  - Implementation . . . 8
- Data Diskette . . . 9
- Organization of the Book . . . 9
- References . . . 10

Chapter 2: Sources of Information for Forecasting . . . 13
- National Water Use Data . . . 13
  - Residential Water Use . . . 14
  - Commercial and Industrial Water Use . . . 15
- Collecting Water Use Data . . . 17

Chapter 2: Sources of Information for Forecasting—*continued*
Organizing Data . . . 17
Time-Series Data . . . 17
Cross-Section Data . . . 18
Pooled Cross-Section and Time-Series Data . . . 18
How Much Data Disaggregation? . . . 18
Other Variables . . . 19
Weather . . . 19
Population . . . 19
Personal Income, Employment, and Production . . . 19
Buildings . . . 20
Preliminary Data Analysis . . . 20
Cleaning the Data . . . 20
Basic Statistical Analysis . . . 20
Confidence Intervals . . . 21
Irregular Distributions . . . 22
Summary . . . 23
References . . . 24

Chapter 3: Curve Fitting . . . 25
Goodness of Fit . . . 26
Regression Analysis . . . 27
Statistical Significance . . . 28
Standard Errors and t Statistics . . . 29
Confidence Interval . . . 29
R^2 and the F Statistics . . . 30
Standard Error of the Estimate . . . 31
Elasticity . . . 31
Unbiased and Reliable Parameter Estimates . . . 31
Serial Correlation of Error Terms . . . 31
Heteroscedasticity . . . 32
Specification Problem . . . 32
Multicollinearity . . . 33
Nonlinear Relationships . . . 33
Other Curve-Fitting Techniques . . . 34
Summary . . . 35
References . . . 35

Chapter 4: Water Use Coefficient Models . . . 37
Trending Per Capita Use Rates . . . 39
Per Capita Forecasts . . . 44
Forecasting With Unit Water Use Coefficients . . . 45
Comparing the Forecasts . . . 46
Divergence Between Forecasts—The Underlying Dynamic . . . 46
Value of Per Customer Forecast Approach . . . 46
Forecast Accuracy . . . 47
Adjusting Forecasts for Price and Conservation Effects . . . 48
Summary . . . 49

Notes 50
References 51

Chapter 5: Causal/Structural Forecast Models 53
Estimating a Causal/Structural Model With Household Data 54
Sample Size 54
Explanatory Variables 55
The Data 55
Exploratory Data Analysis 56
Regression Equation 58
Significance Levels of the Estimated Coefficients 60
Coefficient Confidence Intervals 60
Single-Household Forecast 61
Total Population Forecast 62
Estimating a Causal/Structural Model With Time-Series Data 63
Improving Forecast Accuracy With Disaggregation 65
Slope and Intercept Dummy Variables 66
Cross-Section, Time-Series Models 66
Forecast Evaluation 66
Summary 67
Notes 68
References 71

Chapter 6: Forecasting Seasonal and Peak Water Demands 73
Averaging Procedures 74
A Simple Averaging Technique 74
13-Month, Moving-Average Method 77
Deseasonalizing Water Use 78
Exponential Smoothing 78
Simple Exponential Smoothing Example 79
Choosing a Smoothing Constant 80
Winter's Exponential Model 80
Updating Equations 80
Choosing Smoothing Constants 82
Starting Values 82
Forecasts 82
Regression Approaches 83
Graph of the Seasonal Regression 85
Coefficients of the Seasonal Regression 85
Summary 87
Notes 87
References 88

Chapter 7: Population, Economic, and Technological Forecasts 89
National Population Forecasting 89
Population Trends 91
Regional Population Trends 91

Chapter 7: Population, Economic, and Technological Forecasts—*continued*
Regional Population Forecasts 92
A Checklist 92
Other Demographic Factors 93
Economic Forecasts 93
Accuracy Issues 94
A Checklist for Regional Economic Forecasts 94
Personal Income 95
Effect of Inflation 95
Income Trends 96
Technology Forecasts 96
Technology and Industrial Water Use 97
Appliances 97
Clothes Washers 97
Dishwashers 98
Garbage Disposals 98
Reverse-Osmosis Water Filters 98
Water Softening 98
Evaporative Coolers 98
Summary 99
References 100

Chapter 8: Weather and Climate 101
Correlation of Temperature, Rainfall, and Water Use 102
Correlation Coefficients 102
Multivariate Models 104
Other Weather Variables 106
Individual Weather Variables 106
Potential or Measured Evapotranspiration 106
Cycles 107
Trends 108
Heat-Island Effects 108
Global Warming 109
Summary 109
Notes 109
References 110

Chapter 9: Price Effects 111
Schedules of Rates and Fees 111
Monthly Service Fee and Constant Volume Charge 112
Decreasing Block Rates 112
Increasing Block Rates 113
What Is the "Price" of Water? 113
Unmetered Service 113
Metered Service With Commodity Charges 115
Price Elasticities of Demand Revenue 115
Impact of Metering 117
Controlled Comparisons 118

Econometric Studies . . . 118
Qualitative Findings . . . 119
Quantitative Findings . . . 120
Residential Price Elasticity of Water Demand . . . 120
Industrial Water Price Elasticities . . . 120
Commercial Water Price Elasticities . . . 121
Using Price Elasticities in Demand Forecasts . . . 122
Example Computation . . . 122
Translating the Revenue Objective Into a Rate Increase . . . 122
Selecting the Price Elasticities . . . 123
Estimating Quantity Impacts and the Revenue Target . . . 123
Cautions in Applying Elasticities . . . 124
Summary . . . 125
Notes . . . 125
References . . . 127

Chapter 10: Water Conservation . . . 129
Types of Conservation Programs . . . 129
Evaluating Water Conservation Programs . . . 131
Incorporating Conservation Into the Forecast . . . 132
Synergism, Competition, and Independence . . . 133
Estimating Water-Savings Potential . . . 134
Variability of Conservation Program Impact . . . 134
Water Conservation Goals . . . 134
Estimating Conservation Effects With Local Studies . . . 136
Aggregate Before and After Analysis . . . 137
Control Group Comparisons . . . 137
Example Analysis . . . 138
Setting Up a Controlled Comparison . . . 138
Disaggregated Causal/Structural Model . . . 140
Summary . . . 142
References . . . 143

Chapter 11: Forecast Uses, Evaluation, and Improvement . . . 145
Forecasting Process . . . 145
Short-Term Forecasts . . . 146
Long-Term Forecasts . . . 147
Medium-Term Forecasts . . . 147
Forecast Interval Estimates . . . 148
Regression-Generated Confidence Intervals . . . 148
Monte Carlo Simulations . . . 149
Alternative Scenarios . . . 149
Forecast Evaluation . . . 149
Mean Squared Error (MSE) . . . 149
Mean Absolute Deviation (MAD) . . . 150
Mean Absolute Percent Error (MAPE) . . . 150
Bias . . . 150

Chapter 11: Forecast Uses, Evaluation, and Improvement—*continued*
Error Measures and Loss Functions 151
Strategies to Improve Accuracy 151
Why Forecast Formally? 152
Notes 153
References 154

Appendix A: Direct Data Collection 155
Customer Feedback 155
Expert Opinion 156
Alternative Scenarios 156
Delphi and Other Expert Opinion Techniques 156
Designing a Survey 157
Steps in Designing a Survey 157
Questionnaire Development 158
Reviewing Questions 159
Target Population 159
Nonresponse Bias 159
Selecting the Sample 160
Systematic Sampling 160
Sample Size for a Proportion 160
Mail Survey 162
Telephone Survey 163
An Example Telephone Survey 163
An Alternative Approach 164
Summary 164
References 164

Glossary 165
Additional Sources of Information 171
Index 175

Figures

Figure 2-1 Frequency Distribution of Summer Residential Water Purchases 23
Figure 3-1 Scatterplot Diagram for an Increasing Logarithmic Relationship 26
Figure 4-1 Annual Per Capita Water Use 1980–1993 and Four-Period Moving Average 41
Figure 4-2 Annual Per Capita Water Use 1980–1993, Linear Trend Line 42
Figure 4-3 Annual Per Capita Water Use 1980–1993, Logarithmic Trend Line . . . 43
Figure 4-4 Annual Per Capita Water Use 1980–1993, Power-curve Trend Line . . . 43
Figure 4-5 Annual Per Capita Water Use 1980–1993, Exponential Trend Line . . . 44
Figure 5-1 Frequency Distribution of Residential Water Use Sample 57
Figure 6-1 Monthly Residential Water Use Series 74
Figure 6-2 Monthly Pattern of Water Use in the 1980s 76
Figure 6-3 13-Month, Moving-Average, Seasonal Model 79
Figure 6-4 Exponential Smoothing Model 81
Figure 6-5 Seasonal Regression and Plot of Residuals 84
Figure 7-1 US Population Projection, US Census Bureau 90
Figure 8-1 Monthly Average Temperature 102
Figure 8-2 Monthly Precipitation 103
Figure 8-3 Annual Average Temperature: Los Angeles, Calif., Civic Center 108
Figure 9-1 Fixed Fee and Constant Commodity Charge 112
Figure 9-2 Decreasing Block Rate Schedule With Fixed Fee 114
Figure 9-3 Increasing Block Rate Schedule With Fixed Fee 114
Figure 10-1 Flow Diagram of Goal Achievement Process 136

Tables

Table 1-1	Types of Forecasts and Major Applications	3
Table 1-2	End-use Composition of Single-family Residential Water Demand	7
Table 2-1	Average Per Capita Water Consumption by USEPA Region	14
Table 2-2	Per Capita Water Consumption and Customers Served	15
Table 2-3	Residential Water Use in the Boroughs of New York	15
Table 2-4	Commercial and Institutional Water Use Coefficients	16
Table 2-5	Commercial Water Use Rates by Floor Space of Establishment	16
Table 4-1	Historic and Forecast Water Sales	38
Table 4-2	Water Customers and Population—Historic Data and Forecasts	39
Table 4-3	Water Use Per Customer and Per Capita—Historic Data and Forecasts	40
Table 4-4	Use Rates for Generic Customer Classes That Lead to a Switch in the Direction of Per Capita Water Use	47
Table 5-1	Averages and Standard Deviations of Sample Data	57
Table 5-2	Correlation Matrix for Residential Microdata—Raw and Log-Transformed Versions	58
Table 5-3	Regression Analysis of Sample Residential Water Use Data	59
Table 6-1	Method of Averages for Identification of Seasonal Pattern of Water Use	75
Table 6-2	13-Month, Moving-Average, Seasonal Model	77
Table 6-3	Summary Output: Seasonal Residential Water Use Equation	86
Table 8-1	Summary Output: Monthly Aggregate Residential Water Use Equation	105
Table 9-1	Typical Block Rate Schedules	113
Table 9-2	Water Use in Metered and Flat Rate Areas, Western United States	117
Table 9-3	Controlled Comparison of Water Usage Rates	118
Table 9-4	Estimated Price Elasticities by Region for Selected Price Levels	120
Table 9-5	Estimated Price Elasticities, Results of Demand Studies	121
Table 9-6	Industrial Water Demand Elasticities	121
Table 10-1	Implementation of Water Efficiency Programs in the United States	131
Table 10-2	Engineering End-use Estimates of Water Savings	135

Acknowledgments

The authors are indebted to many theoreticians and practitioners. We are particularly grateful for inspiration from the work of J. Scott Armstrong and the encouragement of Helmut Frank, Lester Taylor, Charles W. Howe, and Andreas Weigend. We appreciate the many students and clients who have helped us develop a deeper understanding of forecasting issues. We are particularly grateful to Donald E. Agthe (who coauthored chapter 9), W. Mark Day, Molly Longstreth, John R. Morris, Mary Goldberg, the late William Bruvold, Anthony F.C. Fisher, and Nile Hatch for discussions of water demand models.

We also thank the following reviewers, whose valuable suggestions helped shape and refine this publication: Steven L. Olson, Rocky D. Wiley, William O. Maddaus, Jack A. Weber, Jeffrey P. Featherstone, Dan Rodrigo, and Benedykt Dziegielweski.

We also wish to thank staff of the water departments of Miami, Fla.; Boston, Mass.; Houston, Texas; Dallas, Texas; Tucson, Ariz.; Denver, Colo.; Seattle, Wash.; and the Metropolitan Water District of Southern California, Los Angeles, Calif., for their assistance in supplying documents and supportive discussion of the problem of forecasting.

Chapter 1

Introduction

Urban water demand forecasting is the process of making predictions about future water use based on a knowledge of historic water use patterns.

An effective water demand forecasting program can help utility officials answer such questions as the following:

- What is the seasonal pattern of water use?
- How large is peak usage likely to be?
- What size treatment plant and water supply system is needed to accommodate peak usage?
- How does weather affect urban water use?
- How can the effectiveness of a water conservation program be evaluated?
- What effect will rate increases have on water sales?

This book looks at these problems from a comprehensive viewpoint, suggesting useful methods of statistical analysis and criteria for key inputs to urban water demand forecasts, such as population, economic trends, and the weather.

Many water utilities rely on simple forecasting techniques, such as using ratios indicating the water use per person in the service area multiplied by population. These methods rely on the validity or accuracy of population forecasts typically produced by other agencies.

Today, many larger utilities analyze water use patterns by customer class, believing that such *sectoral disaggregation* provides better forecasts and more information for planning. This more sophisticated method of analyzing and forecasting urban water demand raises a number of issues about projections of the number of residential, commercial, industrial, and public customers—issues relating ultimately to the basic form and growth or decline of a community or region.

There are many excellent books on forecasting methods (e.g., Pindyck and Rubinfeld 1991; Armstrong 1985; Bails and Peppers 1993; and Makridakis and Wheelwright 1982) but only a few studies of urban water demand forecasting. Listed in the references section at the end of this chapter, some book- or monogram-length treatments of urban water demand include Kindler and Russell 1984; Jones et al. 1984; Boland et al. 1983; Grima 1972; and Linaweaver, Geyer, and Wolff 1966. The emphasis in this book is on applications, rather than theory, and the text is supported by a series of examples presented on the accompanying data diskette.

Technical terms are explained when first introduced. Many important statistical and econometric terms also are defined in the glossary.

The book is designed for engineers and planners with forecasting responsibilities in water utilities. It also should be valuable as a supplemental exposition for water resource analysts, as an addition to community library holdings on municipal planning issues, and as a window into applied work for engineering, planning, and community design students.

Why Forecast?

Virtually all decisions are based on forecasts or predictions about the future. Many times, these forecasts exist only in the mind of the decision maker and reflect a belief that the future will be similar to the past. Although this "same-as-always" forecast is often correct, it is inflexible in the face of changing conditions, such as a large price increase or a drought, and will periodically result in expensive policy blunders.

Many factors influence water use, including the weather, over which utility operators have no influence; population and industrial location, over which they have some limited influence; and price and conservation programs, which are determined or at least strongly influenced by utility officials. Both revenue and sales volume can also be changed somewhat by changes in internal utility operating procedures.

All of these factors signal the value of a carefully executed forecasting program for the water utility.

A successful forecasting process serves management by providing organized information about the past and its implications for the future. One benefit of formal methods is that they encourage the forecaster to make the assumptions of the model explicit, so that the assumptions can be examined. The forecaster also must collect and analyze historic data, thereby avoiding errors based on unsupported beliefs about the past. Careful analysis and forecasts of urban water demand, for example, can improve the timing of expensive system expansions, often saving an urban water utility many times the cost of the forecast.

Urban water demand forecasts are necessary in some cases to meet regulatory requirements. Many states now require that water utilities submit water conservation or efficiency plans. In some instances, targets are prescribed, e.g., "For 1995 attain an average annual water use of less than 160 gal per capita per day (gpcd)." Failure to meet such targets can mean fines and penalties. Under these circumstances, careful inclusion of weather impacts in the demand forecast is critical. Statistical models can also capture the specific effects of various conservation measures, such as installation of water-efficient appliances and plumbing fixtures, lawn size restrictions, and free distribution of water-saving devices.

Major Uses of Forecasts

The physical setup of urban water systems, including wells, pumps, distribution systems, treatment plants, reservoirs, dams, and aqueducts, merge with changing consumer demand to produce specific needs for forecasts. These needs can be classified by time horizon and policy issues. Table 1-1 shows three types of forecasts and their associated major applications. Each of the three types will be examined more closely in following sections.

Table 1-1 Types of forecasts and major applications

Forecast Type	Time Range	Application
Short term	Up to 1 or 2 years	Operations, revenue forecasting
Medium term	Between 1 or 2 and 7 to 10 years	Treatment and distribution investments, setting water rates, conservation programs
Long term	More than 7 to 10 years	Raw water supply, overall capacity

Long-Term Water Demand Forecasts

The challenge of long-term forecasting results from the increasing difficulty of predicting events more and more distant from the present. However, the commitment involved in capacity planning, which depends on the long-term forecast, is very great because urban water systems have capital-to-output ratios rivaling such industries as telecommunications and electric power. Plant and equipment costs for producing a unit of potable water are several times greater than water supply costs for typical manufacturing activities or agriculture. Lead times for expansion also are increasing as permit applications require more preparation, particularly to assess environmental effects.

Forecast errors increase with the length of the forecast period, suggesting that long-run forecasts used for capacity planning may contain large errors. The risk to water utilities from forecast errors is two-sided. On the one hand, water rates and charges can become burdensome as an overbuilt facility comes on line, imposing debt service on a rate base that is smaller than projected. On the other hand, water shortages from inadequate raw water acquisition, treatment, or storage facilities also impose costs to customers. Customer costs take the form of losses in landscaping, convenience of water use, and possible constraints on new construction.

In the years after World War II, water supply risks were usually managed by opting for the largest feasible project. As long as economic and population growth continued, this tactic worked well. However, population growth has slowed in many areas of the nation, with some central urban areas actually experiencing losses in population. Thus, accurate long-term forecasts have become crucial to utility planning.

Medium-Term Water Demand Forecasts

Medium-term forecasts—for 1 to 10 years—are commonly developed for planning improvements to the water distribution and treatment system and for setting water rates. Medium-term forecasts focus on the variability of water consumption by a fixed or slowly changing customer base. Variations in demand over this time period are driven by the variability of the weather and, for industrial and commercial water sales, the business cycle.

Short-Term Water Demand Forecasts

Short-term water demand forecasts support water system operations as well as budgeting and financial management. Forecast horizons are commonly a year or so. Detail over periods as short as a month or even a day can be relevant. Highly accurate forecasts can generally be developed for typical weather conditions because of the short time horizon. Forecast errors in the short run arise primarily from the inherent variability and unpredictability of the weather and human behavior. These effects can be documented and effectively analyzed by short-run forecasting methods.

Major Influences on Water Demand

Several factors have significant impacts on water demand, including population, economic cycles, technology, weather and climate, price, and conservation programs. These factors and others are discussed in the following sections.

Population, Economic Cycles, and Technology

Population growth is often the major trend factor in water use. In the United States, the Southwest and Florida continue to be magnets for migration, attracting those seeking a climate suitable for retirement. Conversely, in many of the older urban areas of the nation, population and employment are declining, implying reduced needs for water in the future.

Business-cycle factors impact water use since fluctuations in industrial and commercial production rates translate into commensurate changes in water demand. Changes in personal income also influence water use because, in the jargon of economics, water is a "normal" good. Its consumption increases, other things being equal, when family income rises. Consumption is likely to drop if the primary breadwinner in the household is out of work.

Technological change can also impact water use over time. Widespread installation of garbage disposals and automatic dishwashers in homes may increase domestic water use, while adoption of drip irrigation may reduce it. New production methods in industry may use either far more or far less water than previous methods, while requirements and opportunities for water reuse may dramatically cut water requirements.

Weather and Climate

Seasonal components in water use are generated primarily by the local climate. Summer peaking demand is typical. Higher summer demand levels are related to water use for outdoor activities, including lawn watering and gardening, and use of evaporative coolers. In some cold climates, there are winter-peaking systems; water flow is used to prevent the freezing of pipes on the coldest days. Seasonal demand patterns are important in planning the capacity of water treatment and distribution systems. Short-term patterns are also critical for scheduling maintenance times for reservoirs, pumps, and mains.

Price

Price effects are important for short-, medium-, and long-term forecasts. Both water use and utility revenue are directly impacted by water rate changes. In the short term of a few months, rate hikes can promote behavioral changes in consumers. These changes may include taking shorter showers or reducing car washing and watering of lawns. In the longer term, if a noticeable rate hike keeps pace with inflation, consumers may adapt through their selection of water-using durable goods. They will tend to favor appliances with lower water use ratings and may consider innovations in landscaping designed to cut back on water use. As the relative cost of water compared to other goods drifts up, perhaps due to the exhaustion of nearby and inexpensive supply sources or burgeoning federal water quality requirements, water price or rate effects are likely to be more noticeable.

Efficiency and Conservation Programs

Water efficiency and conservation programs typically couple an appeal to civic virtue with information on how to use less water. Crisis programs resulting from drought or other supply interruptions have generated large, albeit temporary, reductions in water use. Programs designed to permanently change individual behavior are capable of generating long-term reductions in water usage. Particularly effective are programs that encourage structural changes such as the use of low-flush toilets and low-water-demanding vegetation. Conservation program effects must be thoughtfully included in the water demand forecasting model to minimize errors in projected water use and revenue. Regardless of the time frame of the forecast, an effective conservation program will cause a reduction in water use and a drop in revenue compared to what might be expected without a conservation program in effect.

Other Factors

Physical deterioration in the water distribution system is an often overlooked factor influencing water production requirements. Degradation of the water distribution infrastructure is likely to result in increased losses from leaky and broken pipes. The size of these losses in an older system can be hard to estimate, however, since certain types of water meters may underregister as they age. Thus, some unaccounted-for water may be delivered to customers rather than being lost. End-user water losses may also tend to increase with the age of an urban area. Thus it may be desirable to include estimates of future lost water in a complete water use model. The inclusion of water losses in the forecasting models would also help provide a rational basis for replacement and maintenance policies for pipelines, valves, and meters.

Water Demand Forecasting Approaches and Techniques

A wide range of methods is used for forecasting. The method(s) chosen by a particular utility depends on the technical sophistication of the analyst, the resources devoted to the forecasting process, and the available data. Generally the quality and sophistication of forecasts improve as the utility gains experience and the commitment of utility management to the process increases.

Forecasting is useful, even if the accuracy of long-term forecast models is likely to be low. A conceptually valid water demand model aids in considering the impact of changes in such variables as population, industrial production, climatic conditions, price, and water conservation measures.

Most urban water forecasting approaches use a statistical appraisal of per capita (or per customer) water use rates. Projected water use per capita (or per customer) is multiplied by the population forecast to obtain a forecast of total water use. The greatest differences among forecasting methods involves how the per capita forecast is derived and the degree of disaggregation by customer type.

Per Capita and Other Unit Use Coefficient Approaches

In the 1980s, a US Army Corps of Engineers survey found the majority of urban water systems relied on per capita demand extrapolations, often with trend adjustments (US Army Corps of Engineers 1981). Based on surveys mailed to a sample of American Water Works Association (AWWA) member utilities, similar results applied in 1991.

Larger urban water systems often base demand forecasts on unit water use coefficients calculated for residential, commercial, industrial, and public customers. A variant of the unit-use coefficient approach calibrates the demand forecast to the land-use plan in the utility service area. Residential land is estimated to account for X water use per acre per year, commercial uses for Y water use per acre per year, and so on. The effectiveness of land-use-based water demand forecasts is greatest in those areas with strict land-use regulations, comprehensive land-use planning, and a stable industrial structure.

End-Use Models

End-use model forecasts use extensive detailed information about customer behavior. They build up a forecast from data on the inventory of water-using appliances and fixtures and typical patterns of behavior. For example, the number of laundry cycles run by a typical family in a week may be estimated, along with the appliance-dependent water use per cycle. This allows for the calculation of a per capita per day water use figure for washing machines (see Table 1-2). Note that the values in Table 1-2 were derived to be typical for California. The end-use approach also is applicable to industrial water demand, where engineering techniques, such as linear programming, can be applied to investment decisions regarding water recycling equipment.

Table 1-2 End-use composition of single-family residential water demand

End-use	Percent of Total Use (likely range)	Gallons per Person per Day (likely range)
Indoor Uses		
Toilet flushing	20–25	15–25
Showering/bathing	15–20	10–20
Washing machine	10–15	10–20
Dishwasher	2–5	0–5
Faucet	5–10	5–20
Evaporative cooler	0–2	0–5
Humidifier	0–2	0–5
Leaks and drips	0–5	0–10
Outdoor Uses		
Lawn and garden watering	25–30	25–35
Swimming pool	0–5	0–5
Car washing	0–5	0–5
Driveway cleaning	0–2	0–2

Source: AWWA (1993).

Extrapolation

Extrapolation encompasses a variety of techniques, including simple time trends, exponential smoothing, and Box–Jenkins (autoregressive integrated moving-average) models to project historic water use trends into the future. These models rely on the assertion that future changes in water use can be predicted based on historic changes in water use (while ignoring all other possible influences). These methods can provide reasonably accurate forecasts as long as the future is essentially similar to the past. Accuracy is improved by disaggregating water use by customer class and, if appropriate, through geographic considerations. Since future trends are unlikely to repeat past trends, extrapolation methods are of limited value for long-range forecasting.

Causal/Structural Models

Since the 1960s, economists and social scientists have criticized per capita and unit-use coefficient forecasting methods for ignoring socioeconomic factors. In particular, such fixed coefficients do not reflect the level of water rates, i.e., price effects. This is a problem because water usage responds to price changes and real (inflation-adjusted) water rates do increase from time to time. If trends in water prices, personal income, ownership of water-using appliances, population, urban density, and other factors are to be used together in a forecasting model, causal/structural modeling is appropriate.

Choice of Methods

In a typical application, system water demand is disaggregated, or broken into components, such as residential, commercial, industrial, and public water. Water use for each of these customer classes might be forecast using a different approach or technique. Residential demand, for example, could be forecast with a structural regression model relating overall residential water use to factors like community incomes, the level of water rates, and population growth. Commercial water demand forecasts might be based on water use per customer connection, developed from company records on past usage and extrapolations of business growth. There could be a single large industrial water user in the community that provides the utility with a forecast of its water needs. Public water use could be forecast based on a per capita approach.

Causal/structural models require a lot more data than extrapolation methods, although both require equally accurate data on past water use. Some of the unit-use-rate and extrapolation methods require very little historic data, and in some cases, very little data at all.

Implementation

There are three general options in regard to the basis and organization of the demand forecasts.

1. Borrow from the extensive water demand literature to approximate the most likely response of system water use to anticipated changes in population, the price of water, conservation programs, and other variables. Start with current use rates for the entire water utility or by customer class. Examine changes in these use rates over the past several years and corresponding changes in population, water prices, etc. Develop a representative model using estimates of key parameters from the published literature.

   This option is strongly recommended as a first step in planning discussions, when substantial new investments in the water system are being considered. Current levels of water use per capita and per customer, price and income elasticities from comparable communities, and population and commercial growth projections are relevant inputs to representative models. Rough computations should be encouraged. With some commitment of in-house staff and time, these rudimentary calculations of the effects of anticipated growth, impacts of various proposed rate increases, and assessments of the potential variability of water use based on the weather can be catalogued together in a single spreadsheet to support planning and management discussions.

2. Use what can be called a "template water demand forecasting program," such as IWR-MAIN* (Institute for Water Resources Municipal and Industrial Needs) or WaterPlan,† a computer program commissioned by the California Department of Water Resources.

---

*Available from Institute for Water Resources, US Army, Fort Belvoir, Va.

†Available from California Department of Water Resources, Contact B, DWR Publications Office, (916) 653-1094.

A template water demand forecasting program often uses disaggregated data to provide a framework for water demand forecasting. It may be customized with local coefficients, or used with built-in default values for most parameters, which support its application when little local area data are available.

3. Develop a statistical forecasting model relating water use to the factors influencing water demand. This requires a commitment of resources, but offers the possibility of more accurate forecasts. Such an effort can be helpful when forecast results will be used in helping to make decisions regarding water system investments. Larger water utilities routinely build sophisticated forecasting models, not only because they have larger staffs and budgets, but because the capital at risk with planning decisions is greater.

Data Diskette

Data for the examples included in the text are provided on a diskette so that the reader can reestimate the models and experiment with alternative model specifications. The files on this diskette are in Excel® 5.0 and are also provided as raw data files. In order to fully use the macros and examples in the data diskette, the "Analysis Pak" of statistical and engineering routines must be installed. With this add-on and the ability to write custom routines called *macros* in Visual Basic, Excel 5.0 can act as a powerful statistical tool.

When you start up Excel, open the file CONTROL.XLS on the diskette accompanying this book. This will insert another main menu item called EXAMPLES in the standard Excel menu. By opening EXAMPLES, you will be presented with four options: UNIT USE COEFFICIENT DATA, CAUSAL/STRUCTURAL MODEL DATA, SEASONAL/PEAK DEMAND DATA, and APARTMENT STUDY DATA. These databases are associated with several "buttons" for procedures, such as regression or exponential smoothing, and generally are provided so readers can explore suggestions in the book in a concrete fashion. The data are synthesized from water utility data from Tucson, Ariz.; Denver, Colo.; and samples of typical commercial water use data from a national database.

Organization of the Book

Chapter 2 provides an overview of key features of urban water use in the United States and discusses the types of water use data. Data entry errors, outliers, and other data anomalies are addressed. The chapter also identifies sources where the analyst is likely to find data suitable for developing forecasting models.

Chapter 3 focuses on curve fitting. It explains the basis and requirements of multiple regression. This presentation is designed to facilitate the process of learning how to use regression to develop forecasting models. The chapter also briefly discusses exponential smoothing models.

Chapter 4 considers per capita and unit-use coefficient models. The chapter is based on a sustained discussion of a specific example. It highlights advantages of disaggregating water use data to the level of customer class groupings.

Chapter 5 explains how the analyst can use causal/structural models of water use to relate water use to explanatory factors, such as weather, customer income levels, and water price. An example is provided, showing how a utility could collect data from customers and use this survey data, combined with its own records of water use, to create an effective forecast.

Chapter 6 explores seasonal and peak demand forecasting methods. Simple averaging techniques and decomposition of a data series into trend, seasonal, cyclical, and random components are examined as simple methods of effective seasonality forecasting. Regression methods using a time trend and seasonal dummy variables are explained and illustrated with a comprehensive example.

Chapter 7 presents facts and guidelines pertaining to the use of population, economic, and technological forecasts as inputs to a utility forecasting model. A thorough understanding of these phenomena is critical to all water demand analysis, since the basic "driver" in a demand forecast is usually population. The necessity of adjusting time-series data for inflation is carefully discussed. Trends in population and economic activity are reported and discussed. The changing housing stock, including water-using appliances, is examined.

Chapter 8 considers weather and climate as influences on water use. The chapter discusses weather variables that influence urban water use, briefly touching on evidence for the existence of cycles and long-term climatic trends.

Chapter 9 focuses on water rate and price effects. This chapter provides a summary of the results of more than 40 statistical studies of customer water price responses. A method for using price-elasticity coefficients from the published literature is illustrated. The chapter also provides a nontechnical introduction to the problem of how the price variable should be defined in a forecasting model.

Chapter 10 discusses the integration of conservation and water efficiency programs into water demand forecasts. The methods of estimating conservation from actual field data are examined, including controlled experiments, causal/structural modeling, and end-use analysis.

Chapter 11 considers risk assessment, decision making, and formal confidence intervals for forecasts.

Appendix A presents an introduction to gathering data through random or systematic sampling of utility customers. It includes detailed discussions about how a utility can effectively use a random sample of its customers to obtain information about their water-using fixtures and habits. Appendix A also briefly considers other methods of gathering data, ranging from recurring interactions with utility customers to Delphi techniques.

References

American Water Works Association. 1993. *Evaluating Urban Water Conservation Programs: A Procedures Manual*. Denver, Colo.: AWWA.

Armstrong, J.S. 1985. *Long-Range Forecasting: From Crystal Ball to Computer.* New York: Wiley Interscience.

Bails, D.G., and L.C. Peppers. 1993. *Business Fluctuations: Forecasting Techniques and Applications.* 2nd ed. Englewood Cliffs, N.J.: Prentice Hall.

Boland, J.J., J.L. Pacey, Wai-See Moy, and R.C. Steiner. 1983. *Forecasting Municipal and Industrial Water Use: A Handbook of Methods*, Contract Report 83C-01. US Army Corps of Engineers, Engineer Institute for Water Resources.

Grima, A.P. 1972. *Residential Water Demand: Alternative Choices for Management*. Toronto, Ont.: University of Toronto Press.

Jones, C.V., J.J. Boland, J.E. Crews, C.F. DeKay, and J.R. Morris. 1984. *Municipal Water Demand: Statistical and Management Issues, Studies in Water Policy and Management No. 4*. Boulder, Colo.: Westview Press.

Kindler, J., and C.S. Russell in collaboration with B.T. Bower, I. Gouevsky, D.R. Maidment, and W.R.D. Sewell. 1984. *Modeling Water Demands*. London: Academic Press.

Linaweaver, F.P. Jr., J.C. Geyer, and J.B. Wolff. 1966. *Residential Water Use Project—Final and Summary Report, Report V, Phase Two*. Baltimore, Md.: Johns Hopkins University Press.

Makridakis, S., and S.C. Wheelwright. 1982. *The Handbook of Forecasting: A Manager's Guide*. New York: John Wiley & Sons.

Pindyck, R.S., and D.L. Rubinfeld. 1991. *Econometric Models and Economic Forecasts*. 3rd ed. New York: McGraw-Hill.

US Army Corps of Engineers, Engineer Institute for Water Resources. 1981. *An Assessment of Municipal and Industrial Water Use Forecasting Approaches: Water Conservation and Supply Information Transfer and Analysis Program*, Contract Report 81-CO5.

Chapter 2

Sources of Information for Forecasting

Historic data are essential for reliable water demand forecasts. Utility records are the primary source of this information. How to package the data, how far back in time to go, and what to do when data are incomplete are factors that deserve consideration. After reviewing typical water use rates in the United States, this chapter considers some of these data issues. Differences between time-series and cross-sectional data are defined. Sources of data describing major influences on water use, such as household income, population, and weather, are presented. Disaggregation of water use data is considered with respect to forecasting performance. The discussion suggests approaches to outliers, and reviews basic statistical concepts, such as the mean, median, mode, variance, and standard deviation of a frequency distribution.

National Water Use Data

The American Water Works Association Water Industry Data Base (WIDB) is a collection of information on per capita use, defined as total retail water sales divided by retail population served. Table 2-1 shows how average water sales, measured in gallons per capita per day (gpcd), varied by geographic area during 1990. The areas are administrative regions of the US Environmental Protection Agency (USEPA). The highest per capita water sales—217 gpcd—was in the Rocky Mountain Region VIII, including the states of Colorado, Montana, North Dakota, South Dakota, Utah, and Wyoming. The lowest sales occurred in Region III, in Delaware, Maryland, Pennsylvania, Virginia, West Virginia, and the District of Columbia, at 149 gpcd.

Table 2-1 Average per capita water consumption by USEPA region

USEPA Regions	Gallons Per Capita Per Day (gpcd)
I: Connecticut, Maine, Massachusetts, New Hampshire, Rhode Island, Vermont	153
II: New Jersey, New York	167
II: District of Columbia, Delaware, Maryland, Pennsylvania, Virginia, West Virginia	149
IV: Alabama, Florida, Georgia, Kentucky, Mississippi, North Carolina, South Carolina, Tennessee	158
V: Illinois, Indiana, Michigan, Minnesota, Ohio, Wisconsin	179
VI: Arkansas, Louisiana, New Mexico, Oklahoma, Texas	199
VII: Iowa, Kansas, Missouri, Nebraska	186
VIII: Colorado, Montana, North Dakota, South Dakota, Utah, Wyoming	217
IX: Arizona, California, Hawaii, Nevada	211
X: Alaska, Idaho, Oregon, Washington	196
Average	181.5

Source: AWWA (1991).

Residential Water Use

Water sales can be broken out by customer class. Annual residential water purchases ranged from a low of 70,500 gal per household annually in Region III to a high of 169,800 gal annually in Region IX (Arizona, California, Hawaii, and Nevada). Typically, residential water use is 50 to 65 percent of system water demand. Residential demand tends to make up a larger portion of system water use in utilities serving smaller urban areas. In larger urban areas, a greater proportion of total water is devoted to commercial, industrial, and public sector uses.

Other AWWA data give per capita water use by the number of customers or people in the service area. Here, population refers to retail population served, and water quantity is counted as all deliveries—residential, commercial, industrial, public, and other. Table 2-2 shows that per capita consumption tends to increase with urban size. The common explanation for this is that nonresidential components of system water use increase with urban size, boosting the per capita ratio.

For the nation as a whole, per capita water consumption has increased over time, although the growth rate has slowed recently.

Table 2-2 Per capita water consumption and customers served

Population Served	Gallons Per Capita Per Day (gpcd)
10,000–25,000	158
25,000–50,000	168
50,000–100,000	174
100,000+	183

Source: AWWA (1991).

Many factors contribute to growth in per capita water use, including technological change and migration from urban areas to the suburbs. Consumers in the 1990s use many water-using appliances, equipment, and fixtures, ranging from washing machines and dishwashers to hot tubs, swimming pools, and fountains. Historically, lavish water use has been associated with wealth. As living standards rose after World War II, and owning a house with a variety of water-using appliances and a lawn became a middle class goal, per capita water use surged (Solley et al. 1988).

Most public water supply systems distinguish residential customers from commercial and other customer types for purposes of billing. Water use patterns typically differ between residential and other billing classes, making these distinctions useful to the analyst. In smaller utility service areas, where a few firms dominate a particular category, water use patterns may fluctuate erratically within each category from year to year.

Table 2-3 shows estimated residential water use rates per capita per day for single-family, duplex, and multifamily dwellings in the boroughs of New York City for 1985. These data represent averages for metered and unmetered accounts. Thus, their wide variation in use rates may be due, in part, to water losses through leakage allocated to unmetered accounts. Variations such as this signal the importance to water utilities of developing use rates with data from the local water system.

Table 2-3 Residential water use in the boroughs of New York, in gallons per capita per day (gpcd)

	Bronx	Brooklyn	Manhattan	Queens	Staten Island
Single family	97.6	87.4	165.6	91.0	102.5
Duplex	125.9	92.3	120.3	94.6	95.2
Multiple family	85.2	80.0	132.3	131.5	86.6
Total residential	85.8	83.7	132.4	110.3	97.8

Source: Hazen and Sawyer (1989).

Commercial and Industrial Water Use

Studies show commercial water use to be related to employment or floor space. Table 2-4 shows typical water use coefficients associated with a variety of commercial and institutional categories.

Table 2-4 Commercial and institutional water use coefficients

Category	Water Use Per Employee Per Day, in *gallons*
Hotels and restaurants	186.6
Schools and universities	55.6
Recreational facilities	225.5
Electric and gas utilities	6.7
Food retail	31.3
Barbers	380.2
Hospitals	82.2
Miscellaneous commercial	47.2

Source: Army Institute for Water Resources (1987).

Commercial water sales by type of establishment and floor space for the Denver metropolitan area in 1981 are presented in Table 2-5. Office buildings averaged about 0.09 gal per day (gpd) per square foot of floor space. In each category, establishments with less than 3,000 square feet of floor area had higher than average water consumption per square foot.

Table 2-5 Commercial water use rates by floor space of establishment

Restaurants	0.57
Hotels and motels	0.27
Theaters	0.18
Medical buildings	0.18
Stores	0.13
Financial buildings	0.11
Shopping centers	0.10
Office buildings	0.09
Factories	0.08
Warehouses	0.04

Source: M. Hoddinot, Denver Water Department, data for 1981.

While these use rates provide a place to begin analysis of nonresidential demand categories, they often vary over time and among locations. Thus, local data must be developed when possible. Detailed information about water use rates by customer category can improve water utility forecasting performance (see chapter 4).

Collecting Water Use Data

Utility billing records are the primary source of information about system water use. Detailed information from a metered water system about individual customers and groups of customers provides the analyst with a rich source of data. The data available from an unmetered system are likely to be highly aggregated, preventing analysis of demand for individual categories of users.

Rate schedules applying to customers within the utility service area should be compiled and kept indefinitely, so that the various price/rate proxies discussed in chapter 9 can be identified and analyzed. A detailed history of the implementation of water conservation and water restriction measures also is essential. At least 30 years of historic data should be maintained in as much detail as possible.

In the past, customer billing records typically were purged at relatively frequent intervals, such as every two years, to make room for new data. Today, with computer hard drive sizes reaching 1 gigabyte and laser discs offering almost limitless data storage capabilities, there is no reason to delete billing information.

A wise choice for a utility, therefore, is to record all information on water purchases in a standard format that allows convenient retrieval at any time in the foreseeable future. Maintaining records for each billing period by individual customer allows future analysts to aggregate the water use data in whatever way is desirable.

The question arises, "How many years of data should be collected?" In general more data are better than less data. Careful documentation of the data formats and definitions ensures the usability of stored data in future years. In purely statistical terms, a minimum of 20 years of annual data and 4 to 5 years of monthly data are necessary to support the development of forecasting models. More reliable results can be obtained from these models with 30 to 40 years of annual data and 10 to 15 years of monthly data.

The quantity of water sold or produced by billing period is valuable because water sales or production on monthly, bimonthly, or other short-period cycle can reveal the seasonal pattern of water use. Daily or even hourly use rates are desirable when analyzing key differences between peak and average water use levels.

Organizing Data

Time-Series Data

An historic record of water sales over many days, months, or years is often referred to as *time-series data*. A time series is simply a collection of observations on a particular variable of interest over time. Every time series has a specified interval over which the value of the variable is accumulated (in the case of a flow variable, such as water purchases) or averaged (in the case of a status variable, such as the number of active water meters). The values of status variables are often recorded for specific dates, such as the first day of each month.

Another dimension for classifying data is its level of aggregation. The most detailed and most completely disaggregated data are time series from each customer recorded at close intervals, perhaps hourly, by meters linked by telecommunications with a central computer. More common are data aggregated by customer class, geographic region, and by month or quarter.

Cross-Section Data

Data may also be organized on a *cross-section* basis. Cross-section data consist of a slice of water use patterns at a specific time interval, e.g., recorded water use for individual customers in a district for a single billing period or year. Cross-section data for individual households are frequently the basis of research and analysis published in professional journals. Many of these studies combine cross-section data on water use with survey data on customer characteristics. These data support the statistical estimation of demand functions for residential or other customers, showing the response of water sales to differences in water rates, variation in consumer income, and other influences (see chapter 9).

Cross-section data to support multivariate or structural water demand forecasting models requires random samples of water purchases and data collection focused on characteristics of the customers in the sample. (See appendix A for a discussion of how to take a random sample.)

Within an individual utility, a cross section might consist of a random sample of individual customers by customer category. Water use might also be gathered for groups of customers broken out by pumpage areas, census tracts, or other divisions.

Pooled Cross-Section and Time-Series Data

Some research uses combined time-series and cross-section data, which consist of a number of individual customers or customer groups with water use measured over many days, months, or years. This is usually referred to as *pooled cross-section, time-series data.* Pooled data offer an especially complete type of data for investigating such issues as how water customers in different income groups respond and adapt to higher water rates over time.

How Much Data Disaggregation?

How much data disaggregation should be undertaken for forecasting purposes? An example in chapter 4 shows how forecast accuracy can be improved by separating residential, industrial, commercial, and public water customers.

How useful is disaggregating water use data to fairly specific levels, e.g., residential customers of certain types in certain areas of the water district, and so on? There is no hard and fast rule covering all cases, but the general principles of composition and noise can be applied.

The *principle of composition* provides the impetus to disaggregate demand models into constituent parts. The idea is to separate system demand into component parts and develop explanations for typical water use of customer groupings. This can add precision to forecasts, insofar as the mix of various customer types may change over time in predictable ways.

The *principle of noise* is based on the theory that it is easier to predict behavior in the aggregate than for individuals or smaller groups. This is because of random (capricious) components in behavior, as well as inevitable errors in observation. Ideally, efforts to predict how much a consumer will buy are just as likely to undershoot as overshoot the correct amount. Thus, forecast errors will tend to net out to 0 if enough customer use patterns are being averaged together. This means that we often can predict the water use of several hundred or several thousand customers with a smaller percentage error than the water use of a single or a few score customers. However, predicting the water use of 100 groups of 20 customers and adding the results may provide more accuracy than a single forecast for all 2,000 combined if there are marked dissimilarities among the groups.

One job of water demand forecasters, then, is to identify the appropriate balance between the principles of composition and noise.

Other Variables

A number of demographic, structural, and economic variables are used in making water demand forecasts. The following is a partial list of these variables together with likely sources.

Weather

Weather data are available from local weather stations and published reports of the US Weather Bureau. Additional data may be available from commercial reporting services, research universities, and land grant colleges. The basic weather variables are temperature and rainfall, with these and other variables available on an hourly, daily, or monthly basis.

Population

Historic population data and forecasts are typically published by the US Census Bureau, and promulgated by state and local economic development agencies. Intercensus population figures are often projected using the number of utility customers, school attendance records, and similar indicators. Other demographic characteristics, such as household size, are available in the US census taken every 10 years and may be available from some of the sources mentioned above on a more frequent basis.

Personal Income, Employment, and Production

Income and employment data by community are available from most state economic security agencies, university economic research departments, the US Census Bureau, and on an annual basis from *Sales and Marketing Management** magazine, the US Internal Revenue Service, and the state tax departments. Some banks and state or local government agencies publish economic data on a periodic basis.

Data on commercial and industrial activity for each county in the United States is reported annually by the US Department of Commerce in *County Business Patterns*. More detailed information is available in the census of business every five years and the

*Available from Sales and Marketing Management, New York, N.Y.

decennial census of population and housing. Sources providing income data often provide measures of industrial and commercial activity as well.

Buildings

Total office space or floor area of nonresidential customers also can be useful in making water demand forecasts. Such information typically is available from county tax records. Compilations of real estate data may also be available from the agencies mentioned earlier as well as from industry reports produced by both private and governmental researchers.

While some utilities may view survey work as an unnecessary expense, such activities are comparable to market research conducted by businesses to determine the nature of their customers and their needs.

Preliminary Data Analysis

Cleaning the Data

Examination of the data or, even better, its graphical representation may reveal unusual or seemingly unlikely values, or *outliers*, as well as gaps or missing values.

Outliers can be considered on a case-by-case basis. They may exist as the result of an unusual event. To determine their source, analysts can consult other data sources or discuss the facts with someone familiar with the process involved. Extremely high water use might have been caused by a broken water main or valve. Unusually low measured usage might result from a malfunctioning water meter or the shutdown of part of the supply system, which necessitated temporary water rationing or emergency conservation measures. Data entry is another fertile source of outliers in records of water sales and production.

If an outlier cannot be explained, one option is replacing it by a more reasonable value. If the datapoint comes from a cross-section sample and the object is to estimate a regression, replacing the outlier by the average of the remaining, apparently valid, values is relatively safe. If the datapoint is a data input error, substituting the average of the remaining values will not change the overall average of the sample. Eliminating an unusually large or small datapoint from time-series data can be trickier, since often one wants a "smoothed" value that will not disrupt the seasonal pattern or the trend. In this case, interpolation is a useful option.

Missing values also can be dealt with by such procedures, i.e., replacement by the average value of a series or segment of the data or by an interpolated value. Omitting the entire observation containing the missing value is another option.

Basic Statistical Analysis

After gathering the water quantity data, the next step is to compute basic descriptive statistics. These include means or averages, medians, and modal values, along with variances, standard deviations, and frequency distributions. These terms are described in the following paragraphs.

The *mean* is the arithmetic average of all users. It is obtained by simply adding up all individual uses and dividing by the total number of customers.

Median water use is that of the customer in the middle—half the customers use less water and the other half use more.

The *mode* or *modal value* describes the water use volume category with the highest frequency.

The *variance* is a measure of dispersion around the "central tendency" or mean value of a variable. The variance is

$$\text{VAR} = \frac{1}{N} \sum_{i=1}^{N} (Q_i - \overline{Q})^2 \tag{2-1}$$

Where:

Q_i = quantity of water sold in a billing period to the ith customer, $i = 1, \ldots, N$

Q = mean water use

The *standard deviation* is the square root of the variance. It can be interpreted as the typical difference between an individual observation and the average of the data series. The standard deviation is particularly important when the data are normally distributed. The normal distribution is often referred to as the *bell-shaped curve,* since it is distributed symmetrically around its average value with roughly the shape of a bell. The characteristics of a normal distribution are entirely summarized or described by its mean and standard deviation. When a variable is normally distributed, for example, an additional observation has a 99 percent chance of being within ± 2.57 standard deviations from its average value and a 95 percent chance of being within ± 1.96 standard deviations.

A *frequency distribution* shows the distribution of responses to a question, often in the form of a display. For a variable that can take on a large number of different values, the construction of a frequency distribution requires sorting the data observations into categories. For example, a frequency distribution of water use might summarize individual customer use into categories, such as 0–5,000 gal, 6,000–10,000 gal, etc.

Confidence Intervals

A *confidence interval* shows the statistical accuracy of an estimated population parameter, such as a mean. A 95 percent confidence interval shows the range within which the mean value of a sample drawn from some population is expected to fall 95 percent of the time. If 100 individual samples were drawn from some population, the average value from 95 of them would be within the 95 percent confidence interval of the true but unknown population mean.

The confidence interval for an average is

$$\overline{Q} \pm Z \frac{S}{\sqrt{N}} \tag{2-2}$$

Where:

$\overline{Q}$ = mean water use

Z = is the number of standard deviations corresponding to the chosen significance level from a normal distribution table. For small samples, Z is replaced by the

t statistic from a table of the t distribution (found in the back of most statistics books).

S = standard deviation computed from a random sample of the population

N = sample size

The Z for a 95 percent confidence interval is 1.96. Thus, the 95 percent confidence interval for a sample of N = 200, with estimated $\overline{Q}$ = 11,200 gal, and S = 2,300 gal is

$$11{,}200 \pm 1.96 \times \frac{2{,}300}{\sqrt{200}}$$

which is

$$10{,}881 \leq \overline{Q} \leq 11{,}519$$

Irregular Distributions

The mean, median, and mode are measures of central tendency, and they can indicate other shape characteristics of a distribution. Figure 2-1, for example, shows the histogram or frequency distribution of water sales to single-family residences in a meter-reading section of a large city. The quantity of water sold, recorded over a two-month billing period in late summer, is measured by 5,000-gal increments on the horizontal axis. Each vertical bar shows the number of residences for which water use fell in that particular use category.

Mean or average water use per residential customer is approximately 42,300 gal, while the median consumption is 38,000 gal. The fact that the mean is greater than the median indicates a positively skewed or "left steep" distribution. (Skewness is a measure of the extent to which individual observations are not uniformly distributed around the mean value.) Thus, this distribution does not have the symmetric, bell shape of the normal distribution. However, the distribution in Figure 2-1 is well described by a *lognormal distribution*, except for the highest consumption levels. A distribution is lognormal when the logarithms (base 10) of the observed values are normally distributed. Lognormal distributions are typical for wealth-related variables.*

*Lothar Sachs, in his *Applied Statistics: A Handbook of Techniques*, translated from the German by Zenon Reynarowych, New York, Springer-Verlag, 1984, writes, "A lognormal distribution ... results ... when many random quantities cooperate multiplicatively so that the effect of a random change is in every case proportional to the previous value of the quantity ... the lognormal distribution predominates, in particular, for economic and biological attributes .... Such characteristics in humans include body height (of children), size of heart, chest girth measurement, pulse frequency, systolic and diastolic blood pressure .... Economic statistics with lognormal distributions include gross monthly earnings of employees, turnovers of businesses, and acreages of cultivation of various fruits" (pp. 107–108).

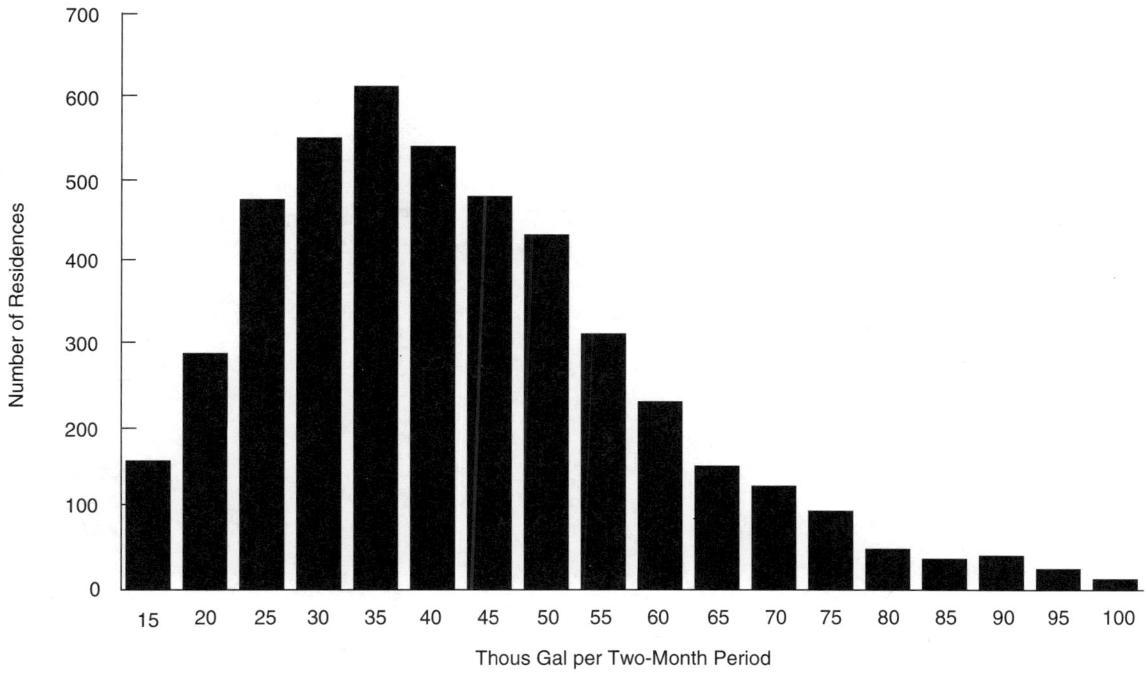

Figure 2-1 Frequency distribution of summer residential water purchases

Summary

This chapter examined typical data on water use rates and discussed key data issues.

Water use per capita, measured as the ratio of total water sales to total population, varied between 149 and 217 gpcd during 1990. Per capita use varied by the size and composition of the community, with larger communities showing higher use rates. Typically, residential water use is 50 to 65 percent of system water demand.

Local data collection is essential for effective water demand forecasting. Water demand analysts must ensure that water use records are suitably archived, aiming for 20 years or more data on water use, preferably in a monthly series. A balance must be struck between detailed descriptions of water use by customer grouping and the stability of water use patterns, which can be determined for larger customer groupings.

The chapter highlighted the basic types of data, such as cross-sectional and time-series data. It identified basic descriptive statistics, such as the mean, mode, median, variance, standard deviation, and skewness. It also considered briefly the problem of missing values and outliers in data.

Finally, the chapter looked at sources of data for weather, population, personal income, employment, production, and commercial and industrial activity.

The underlying purpose of this chapter is to show that having a good understanding of the data used for analysis helps forecasters keep their feet on the ground and avoid errors in analysis.

References

American Water Works Association. 1991. *Water Industry Data Base: Utility Profiles*. Denver, Colo.: AWWA.

Army Institute for Water Resources. 1987. *IWR-MAIN Water Use Forecasting System, Version 5.1, Users Manual and System Description*. Fort Belvoir, Va.

Hazen and Sawyer. 1989. Study of Water Demands on New York City System, Final Report, November 1989. Prepared for the New York State Department of Environmental Conservation and New York City Department of Environmental Protection.

Solley, W.B. et al. 1988. Estimated Use of Water in the United States in 1990. *US Geological Survey Circular 1004*.

Chapter 3

Curve Fitting

Curve fitting is the process of finding a mathematical or statistical function with the same pattern as the data being analyzed. This function can then be used to compute forecasts.

The first step in curve fitting usually is to graph the data. Figure 3-1 shows a scatter-plot diagram for a logarithmic curve, in which some randomness or noise has been introduced. The noise causes the plotted points to depart from the true mathematical form. The idea in curve fitting is to draw a connected curve through this cloud of points so that the "fit" is good.

Sometimes the curve is designed to capture a pattern that appears over time. Thus, aggregate system water use often shows (1) seasonal variation, (2) underlying trends extending over several years, and (3) long-term cyclical behavior related to economic or weather cycles. Time-series techniques, such as exponential smoothing, make it possible to relate future water use to past water use by identifying typical seasonal patterns (seasonality) in water and time trends. An example of exponential smoothing is discussed in chapter 6.

Most civil engineers and water resource analysts are familiar with several types of trend lines, such as logarithmic and exponential curves. A straight-line trend is a relative rarity; usually growth in a time series involves gradual acceleration or slowing. Logarithmic and exponential curves, developed in chapter 4, are often fit to historic data on water use to identify trends.

This chapter focuses on linear regression as a curve-fitting technique. It is intended as a refresher or introduction to statistical concepts used by regression packages, including such business spreadsheet programs as Excel® and Lotus 1-2-3.® The chapter considers regression diagnostic criteria, such as the coefficient of determination or R^2, the F statistic, and the standard errors and t statistics. It develops simple arithmetic examples for computing confidence intervals for the estimated regression coefficients. It discusses violations of the assumptions of the classical linear regression model that can introduce bias or unreliability into parameter estimates—conditions with names such as heteroscedasticity, autocorrelation, and multicollinearity.

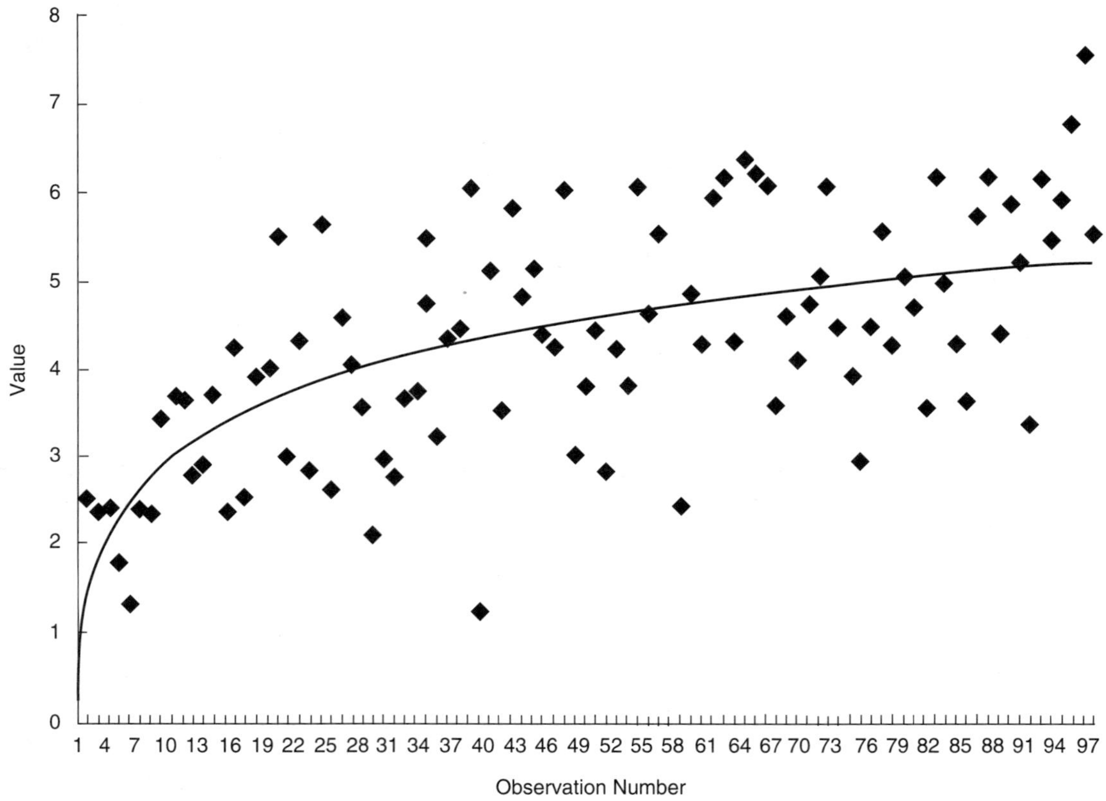

Figure 3-1 Scatterplot diagram for an increasing logarithmic relationship*

**The data shown follow the form $Y = a + b \times ln + E$.*

Where:

- Y = the vertical axis value
- a and b = the parameters describing the specific relationship between Y and $ln(X)$
- X = the horizontal axis
- ln = the natural logarithm to the base e, and the relationship between Y and $ln(X)$ is linear
- E = random component

Also discussed in this chapter are how regression specifications expressed in multiplicative form can be transformed by means of logarithmic transformations into a linear form. Throughout the discussion, various rules of thumb or useful procedures are suggested.

Goodness of Fit

Most methods of curve fitting measure goodness of fit by reference to a summation of squared errors. The curve predicts a certain value for each case in the sample. The sample can be a time series, including observations on water use over time; or the sample can be a cross section of observations on the water use of, for example, various customers during a specific billing period. In any case, this sum of squared errors is usually averaged by dividing by the number of cases or time periods, to create the mean square error or MSE.

$$\text{MSE} = (1/N) \sum_{i=1}^{N} (Q_i - \hat{Q}_i)^2 \tag{3-1}$$

Where:

N = the total number of months in the sample

i = the sequence number of each individual month

Q_i = the observed water use during the ith month

$\hat{Q}_i$ = the estimated water use during the ith month

The MSE is 0 if there is a perfect fit between the estimating curve and the sample observations. Positive and negative errors of the same magnitude are accorded the same weight in the MSE. Because the errors are squared, one large error is given more weight than several small errors adding to the same total amount.

One objective of curve fitting is to minimize this sum of squared errors between the actual data values and an estimated curve.

Regression Analysis

Regression can incorporate the effects of many independent or explanatory variables in explaining the variations in a dependent or target variable. A regression relationship for monthly water use of the average residential customer, for example, might be written as

$$Q = \beta_0 + \beta_1 \, RAIN + \beta_2 \, TEMP + \beta_3 \, PRICE + \beta_4 \, INCOME + \beta_5 \, HHSIZE + \varepsilon \tag{3-2}$$

Where:

Q = water use for each month, in thous-gal units

β_o = the constant or intercept of the model

β_i, i = 1,2,3,4,5 = estimated slope coefficients

$RAIN$ = monthly rainfall, in inches

$TEMP$ = average monthly temperature, in °F

$PRICE$ = the price per thous gal of water, in dollars

$INCOME$ = Average annual income per household, in dollars

$HHSIZE$ = number of persons in household

ε = the residual error of the estimate

The computer program used for solving the linear regression problem determines the β_i values so that the sum of squared residual errors of the estimate for this relationship is minimized.* This equation says that the *true* water use is equal to β_o plus a weighted linear sum of the values of *RAIN*, *TEMP*, *PRICE*, *INCOME*, and *HHSIZE*, plus a random error or disturbance term.

The variable on the left-hand side of the equation is the target variable we are attempting to explain and is referred to as the *dependent variable*. The variables on the right-hand side of the equation are called the independent, explanatory, exogenous, or regressor variables. Taken together, these explanatory variables are called the *specification of the model*. The literature also refers to *specifying variables* in a regression, which means that these variables are the explanatory variables.

The coefficient for each variable shows the change in the dependent variable for a one-unit change in that independent variable, holding all other independent variables constant. Each regression coefficient can be thought of as the partial derivative of the dependent variable with respect to that specific explanatory variable. For example, a coefficient of –3.0 on the independent variable *PRICE* means that a $1.00 increase in the price of water reduces water use by 3,000 gal/month. The size of the coefficient depends on the way the variable is defined. Thus, changing the unit of measurement of price from dollars to cents simply reduces the price coefficient to –0.03, but does not change the result.

The independent variables of a regression equation are seldom able to include all influences on the dependent variable. The effects of numerous possible explanatory variables, each of which is too minor to include in the specification of the regression, or for which there are no usable data, make up the residual or error term. Possible errors in measuring the variables of the model also contribute to the error term.

These considerations are the basis for the assumption that the error process of a regression is generated by or follows a *normal probability distribution*. The normal probability distribution is the famous bell-shaped curve. The underlying principal is the Central Limit theorem, which asserts that the sum of a sufficient number of random variables (almost always) has a probability distribution approximated by a normal distribution. Thus, measurement errors associated with tabulating customer water use will tend to be normally distributed around a 0 mean. Similarly, the combined effect of innumerable small influences omitted from the regression specification will tend to be described by a normal probability distribution.

Statistical Significance

Spreadsheet regression programs and econometrics packages typically provide several diagnostic tests indicating the statistical significance of a regression. Analysis of the residual error term provides a basis for these tests. Typical diagnostic tests include the following:

- standard errors and t statistics for estimated coefficients
- F statistic for the significance of the regression as a whole

---

*The classical linear regression model assumes that the true relationship between a dependent variable and explanatory variables is linear in the parameters to be estimated. Given this and other assumptions, many of which will be discussed later, regression coefficients will be consistent, i.e., the estimated coefficients will converge to the true, unbiased values when large samples are used (see Jones and Morris 1984).

- R^2 and standard error of the estimate describing goodness of fit

Standard Errors and t *Statistics*

The *standard error* for an estimated coefficient provides an indication of the difference between the estimated β value and the unknown true value. The t *statistic* is the coefficient divided by the standard error. Thus, if the standard error for the *PRICE* coefficient in Eq 3-2 is 1.3 when the estimated coefficient is -3.0, then the *t* statistic is -3.0/1.3 = -2.30. (The estimated coefficient and its *t* statistic always have the same sign.) As a rule of thumb, a coefficient is likely to be significant if the absolute value of the *t* statistic is 2 or larger, or equivalently, if the estimated coefficient is at least twice as large as its standard error.

Variables with insignificant coefficients are often dropped from the model.

An exact statement of significance can be obtained from a table of values for the *t* distribution. The significance level depends on both the computed *t* statistic and the *degrees of freedom* of the estimated regression. Degrees of freedom is equal to the number of observations in the data set minus the number of estimated coefficients including the constant.

In the example of Eq 3-2, there are six estimated beta coefficients. Assuming 48 observations in the data set, this leaves 48 – 6 = 42 degrees of freedom. From the table of values for the *t* distribution, found in the back of most statistics books, the critical value for 95 percent significance for a two-tailed test with 40 observations is 2.021. At the 98 percent level, the critical value of the *t* statistic is 2.423. Thus, we could reject the null hypothesis that this coefficient is actually equal to 0 if we accept a 5 percent chance of being wrong. The null hypothesis, however, cannot be rejected if we demand a higher standard of only a 2 percent chance.

A two-tailed test is used to test the null hypothesis that the coefficient is equal to 0 against the alternative hypothesis that the coefficient is not 0 but could be either positive or negative in value. Statistical tests of significance require the statement of a null hypothesis and an alternative hypothesis. A statistic is then computed that results in either being unable to reject the null hypothesis (in effect accepting it as true) or rejecting the null hypothesis (in effect accepting the alternative as true).

For the coefficient on price in Eq 3-2, however, the null hypothesis is that the coefficient is 0 or larger. The alternative hypothesis is that the coefficient is less than 0—a negative value (higher prices reduce water use). Thus, a one-tailed test would be used, making the critical value for a 95 percent level (all in one tail of the *t* distribution) 1.684. In this example, therefore, the critical *t* from the table of 2.021 represents the 0.975 level—a 1 in 25 chance of rejecting a true null hypothesis.

Confidence Interval

A *confidence interval* is an interval around a coefficient estimate that, at a given probability, can be said to contain the true value of the parameter. Assume a 95 percent confidence interval is to be constructed for an estimated coefficient β with a standard error of *SE*. A 95 percent confidence interval is said to contain the true value of the estimated coefficient with a 19 in 20 odds ratio, i.e., 95 percent of the time. Then, if enough observations (more than 120) have been used to estimate the regression, a table of probabilities for the t distribution indicates a critical value of the t statistic of 1.96 for a 95 percent confidence interval. Therefore, the confidence interval will be

$$\hat{\beta} - 1.96\,SE \leq \beta \leq \hat{\beta} + 1.96\,SE \qquad (3\text{-}3)$$

where β is the true value of the coefficient.

For example, if the estimated coefficient of *TEMP* in Eq 3-2 is 0.071, with a standard error of 0.024, and using the critical t statistic of 1.96, the confidence interval is

$$0.071 - 1.96 \times 0.024 \leq \beta \leq 0.071 + 1.96 \times 0.024$$

which is $0.024 < \beta < 0.118$.

However, this assumes a large sample. If the sample degrees of freedom is 42, the critical value from the t table is 2.021 and the confidence interval becomes

$$0.0225 < \beta < 0.1195$$

R^2 and the F *Statistics*

Is the equation itself significant? The F statistic shows whether or not the entire equation is significantly better than the mean value of the dependent variable in explaining water use. An equation would generally be accepted if it is significant at the 0.05 level or better, meaning that there is a 5 percent chance that the equation does not explain variations in the dependent variable better than its mean value. Since the F statistic is adjusted for degrees of freedom, dropping variables with low t statistics from the equation may improve it and make an otherwise unacceptable equation usable.

The R^2 shows the proportion of variation of actual from mean water use explained by the model. For a cross-section regression of individual customers, the R^2 is typically between 0.20 and 0.50. Although an R^2 closer to 1 is better, the R^2 for a cross-section study using individual customers seldom exceeds 0.5.

The adjusted R^2 is a more sophisticated measure of how effective the equation is, and should be used instead of the simple R^2 when equations with different numbers of independent variables are being compared.

The change in the R^2 and F statistics when an independent variable is added to or dropped from the model provides one possible measure of the importance of that variable to the model specification.

While the simple R^2 always increases when additional independent variables are added to the equation, the adjusted R^2 increases when a new independent variable is added only if the gain from the new variable is greater than the statistical cost in degrees of freedom lost. This typically limits the number of variables accepted in a final model to less than 10 and often to less than 5. If the F and R^2 statistics are not satisfactory, some important influence on water use may have been omitted from the model, or the amount of data used in the analysis may be inadequate. It is also possible that there is no significant relationship among the selected variables or that the actual relationship changed midway through the period used for analysis. Graphs of the data may suggest possible changes in the model.

Standard Error of the Estimate

Is the equation sufficiently accurate? The standard error of the estimate provides a measure of the typical difference between the true value of the dependent variable (water use) and the value estimated by the regression. If this value is unacceptably large, the equation needs further work.

Elasticity

An elasticity, such as the price or income elasticity, measures the effect of a change in an independent variable on the dependent variable. Thus, it shows the percentage change in the dependent variable for a 1 percent change in the independent variable. For example, an income elasticity of +0.2 means that for each 1 percent increase in income, the quantity of water purchased rises by one fifth of 1 percent. The income elasticity of demand in a linear regression model (such as Eq 3-2) is

$$\frac{\partial Q}{\partial I} \times \frac{I}{Q}$$

Where:

$\frac{\partial Q}{\partial I}$ = partial derivative of the regression model with respect to *INCOME*, which in a linear model is the slope or coefficient of the *INCOME* term.

I = average income

Q = average quantity

Unbiased and Reliable Parameter Estimates

One objective of conventional regression analysis is to produce unbiased estimates of parameter values when a "true" relationship, such as Eq 3-2, exists.

Consistency of the parameter estimate is a weaker, but related and perhaps more realistic condition. A parameter estimate is said to be consistent when its expected value becomes closer and closer to the true value as the number of observations in the regression increases.

The basic condition for unbiased estimates of regression coefficients and their standard errors is that the error or disturbance components of the regression be normally distributed, independent random variables with 0 mean and constant variance. Several statistical tests are available to examine the error series created by a regression model for violations of these assumptions.

Serial Correlation of Error Terms

An important source of bias often occurs with time-series data, such as monthly water use. With these data, there is a tendency for large positive residuals of a regression in one time period to be followed by large positive residuals in the next time period (or large negative errors followed by more large negative errors). This is called *serial correlation* or *autocorrelation*. Its presence often indicates that some important variable has been omitted from the model specification.

The primary impact of serial correlation of the error terms is on the estimates of the standard errors of the coefficients. The resulting distortion of the size of the standard errors or t statistics may generate results from the computer program showing that some explanatory variables are not statistically significant, when they in fact are, or vice versa. Possible autocorrelation of the residuals can often be detected from an examination of a plot of the residuals. Most statistical computer packages also provide the Durbin–Watson test statistic for first order autocorrelation. The ideal value for this statistic is 2, indicating no autocorrelation. As a rule of thumb, values of this statistic between 1.5 and 2.5 are acceptable, while values outside this range suggest remedial action. Exact critical values can be obtained from the table of Durbin–Watson statistics in any econometrics book.

The best solution for autocorrelation of the regression equation residuals is to "discover" the missing variable and add it to the model specification. Another tactic is to estimate the change or first difference of the dependent variable instead of the actual level. For example, Eq 3-2 could be set up to estimate the change in water use by computing the first differences of each of the variables and using these first differences as the data for the regression. The analyst using such variables should expect a much lower R^2 because a small error in total water use becomes a much larger error in the change in water use. When these tactics are not successful, perhaps because the analyst cannot find data for the needed variable or a first difference would not make any sense, several mechanical fixes are offered in econometrics and forecasting computer programs. Cochrane–Orcutt and Hildreth–Lu are two frequently used procedures.

Heteroscedasticity

Heteroscedasticity is the situation in which the variance of the error process is not constant, but is instead related to the size of the dependent variable or to some explanatory variable(s), such as income. This occurs most often in the analysis of a cross-section sample of individual utilities or businesses of different sizes. Again, the major impact is on the estimates of the standard errors of coefficients. Heteroscedasticity can often be eliminated by converting the dependent and/or independent variable(s) from an aggregate form to a per-capita, per-employee, or per-square-foot basis. For a cross-section study of water utility districts, for example, employing a dependent variable in the form of water use per capita rather than total water use might alleviate problems of heteroscedasticity in a regression relationship.

Specification Problem

The best forecasts are produced when the major explanatory or exogenous variables influencing the target or dependent variable are in the regression equation and insignificant variables are omitted. Some explanatory variables almost always must be omitted from the regression specification, either because the concept they represent cannot be measured or because their effect is not very significant.

Leaving an explanatory variable out of the regression can bias estimated coefficients if the omitted variable is correlated with an explanatory variable in the regression specification. Because the included variable is correlated with an omitted variable, its estimated coefficient picks up the influence that rightfully should be attributed to the excluded variable. The amount of bias is related to (1) the correlation between the omitted and included variable(s), and (2) the magnitude of the true coefficient of the omitted variable.

Weather variables often are not correlated with many of the other significant influences on water use, such as income. This suggests that it may be possible to evaluate the contribution of weather to water demand without developing a full specification of socioeconomic variables in the regression equation.

Multicollinearity

Multicollinearity can exist when two or more explanatory variables in the sample are highly correlated. As a result, regression procedures may not be able to distinguish the separate contributions of these variables to the dependent variable. As a consequence, the estimated regression coefficients can be unreliable.

The extreme case is that of pure or perfect collinearity. If one explanatory variable is three times the other in every case or observation in the dataset, the variables are perfectly collinear. The variables are also perfectly collinear if the sum of several variables is exactly equal to another variable. Then, it becomes impossible to compute the regression.

Ordinary multicollinearity is the situation in which there is a close, but not perfect, linear relationship between some of the explanatory variables in the sample data.

Multicollinearity is usually considered to be a data or sample problem. Thus, the standard recommendation is to collect more data in hopes that additional observations will dispel the high correlations. Other tactics include reducing the dimensionality of the data with principal components. Such techniques may reduce forecast errors, but also introduce bias to the parameter estimates.

One interesting case is when strong correlations exist between variables in the real world, such as between income and wealth. In this situation, regression forecasts can be relatively unaffected by the distortions introduced by multicollinearity if the variables in question remain in the same relationship over the forecast period; one beta will pick up the effect missed by the other.

The principle of *parsimony* (using the simpler model when greater complexity does not provide significant benefits), however, suggests that when two or more variables are highly correlated, one of them should be omitted from the model specification. Thus, most models include either personal income or home values, but not both.

When two of the independent variables are multicollinear, dropping one of them from the estimated model tends to cause large changes in the estimated coefficient and t statistic of the remaining variable. Thus, the appearance of unexpected signs on estimated coefficients and insignificant t statistics often signals multicollinearity.

Nonlinear Relationships

The forecast equation need not be the simple linear form of Eq 3-2. In order to apply standard regression procedures, however, it must be possible to transform the forecast equation into a linear form. There are several ways to do this. A number of water demand models, including the often-cited Howe and Linaweaver (1967) study and variants of the IWR-MAIN model, present demand or water use as the *product* of a number of influences, such as expressed in the equation

$$Q_i = \beta_o \, RAIN^{\beta_1} \, TEMPERATURE^{\beta_2} \, PRICE^{\beta_3} \, INCOME^{\beta_4} \, HHSIZE^{\beta_5} \tag{3-4}$$

where variables are as previously defined.

Written this way, as a multiplicative form, it is not always easy to see why the parameters (the betas) are essentially linear. However, taking the natural logarithms of both sides of Eq 3-3 yields

$$ln(Q_i) = ln(\beta_o) + \beta_1\, ln(RAIN) + \beta_2\, ln(TEMPERATURE) + \beta_3\, ln(PRICE) + \beta_4\, ln(INCOME) + \beta_5\, ln(HHSIZE) \quad (3\text{-}5)$$

Where:

ln = the natural logarithm

Other variables are as defined previously

Eq 3-5 is just like Eq 3-2, except the data on the explanatory and dependent variables are subject to a logarithmic transformation. Eq 3-5 is linear in the parameters β_i and can be estimated with ordinary regression.

The payoff in transforming to an estimating equation that is linear in the parameters is that the well-known formulas and procedures of regression analysis can be applied. By contrast, when the forecast relationship is nonlinear in the parameters, methods of successive approximation must be used to estimate the values of parameters. Such approximation methods are considerably more complex from a mathematical standpoint.

Other Curve-Fitting Techniques

Seasonality can be identified using averaging methods. These essentially take the ratio of the dependent variable for each month to its total for the year. The ratios are computed for several years and then averaged for each month. This provides estimates of the proportion of total water sales typically occurring each month. An example of this method is provided in chapter 6.

Both seasonality and trend can be identified with relatively simple exponential smoothing models or with the more theoretically complex Box–Jenkins models. Neither of these models has been widely used in urban water demand forecasting, although both are suitable for short-term forecasting efforts. The simplicity of use of the exponential smoothing models should encourage their use, while the complexity of Box–Jenkins models may continue to limit their application.

Exponential smoothing models can be estimated with relatively short data series and are adaptive to changes in either trends or seasonality. As new observations are added, the model parameters adjust to fit data from relationships that are changing, as is reasonably the case with a growing urban water system. Chapter 6 presents a complete example of how an exponential smoothing model might be used.

Summary

A variety of techniques is available for curve fitting. They range from simply plotting the data against time and estimating a curve into the future, to extrapolation models that rely entirely on the past values of the dependent variable, to multiple regression that allows formulation of multivariate forecasting models. Regression is often the technique of choice because of its simplicity of use and its desirable statistical properties, which have been widely studied. Additionally, many nonlinear forms used for forecasting can be transformed into a linear form, which can then be directly estimated by multiple regression.

Regardless of the technique used, effective analysis requires a thorough understanding of the variable being forecast and its surrounding environment. Sophisticated methods, while definitely useful, are no substitute for sound judgment in formulating and evaluating models and methods.

References

Howe, C.W., and F.P. Linaweaver Jr. 1967. The Impact of Residential Water Demand and Its Relation to System Design and Price Structure. *Water Resources Research*, 3(1):13–32.

Jones, C.V., and J.R. Morris. 1984. Instrumental Price Estimates and Residential Water Demand. *Water Resources Research*, 20(2):197–202.

Chapter 4

Water Use Coefficient Models

Water use coefficient models are widely applied in forecasting future water use. Per capita water use coefficient models can be used to develop water demand forecasts by multiplying a water use rate by the projected population served by the water system.

A more sophisticated approach calculates and projects water use for each of several relatively homogeneous groups of customers, e.g., single family, multifamily, commercial, industrial, and public (government). Use rates for these customer groupings often are referred to as customer water use coefficients.

Projections of customer water use coefficients are multiplied by estimated future numbers of water customers to produce sectoral water demand forecasts. Such forecasts for each customer sector are combined to arrive at estimates of total system water demand.

Key issues relating to forecasts based on water use coefficients include

- the degree of disaggregation into customer groupings that is desirable
- how to deal with trends
- how to adjust for changes in water rates and conservation practices

The best way to illustrate these issues is with an example.* Accordingly, Tables 4-1, 4-2, and 4-3 will be used in this chapter to present historic data and projections based on both the per capita and per customer approaches. This offers an opportunity to study ways in which the per capita and per customer forecast approaches parallel one another, and how they can differ with respect to the forecasts generated. The growth rates and trends in this example also provide an opportunity to speculate on the underlying socioeconomics, such as the effects of trends in household size and increased efficiency in industrial water use.

---

*Data used in the example was supplied by Nile Hatch, University of California at Berkeley; and Eva Optiz, Planning and Management Consultants.

Table 4-1 Historic and forecast water sales, in *thous gal*

Historic Water Sales					
Year	Residential	Industrial	Commercial	Public	Total
1982	2,680,100	835,898	1,046,490	160,469	4,722,956
1983	2,008,913	746,487	1,139,081	171,910	4,066,391
1984	2,775,236	679,495	1,111,251	147,375	4,713,357
1985	2,987,312	595,199	1,111,099	147,155	4,840,764
1986	2,364,522	817,452	1,162,740	206,003	4,550,717
1987	3,011,218	570,225	1,156,372	157,054	4,894,869
1988	2,674,316	644,913	1,129,876	159,342	4,608,446
1989	2,971,227	602,465	1,127,106	207,493	4,908,291
1990	2,523,795	742,320	1,158,819	189,455	4,614,389
1991	3,337,677	633,325	1,197,257	207,566	5,375,824
1992	2,755,267	678,486	1,111,797	183,827	4,729,377
1993	2,960,973	628,560	1,134,100	156,142	4,879,776
1994	2,692,760	742,446	1,145,441	161,010	4,741,657
1995	3,005,541	810,959	1,156,896	179,079	5,152,476

Forecast Water Sales						
Year	Residential	Industrial	Commercial	Public	Annual Total of Sectoral	Per Capita
1996	2,908,395	715,065	1,183,001	181,128	4,987,589	4,959,834
1997	2,915,690	717,198	1,195,527	180,011	5,008,426	5,022,713
1998	2,922,550	719,303	1,208,184	178,879	5,028,916	5,086,368
1999	2,928,959	721,377	1,220,975	177,733	5,049,045	5,150,809
2000	2,934.904	723,422	1,233,901	176,571	5,068,798	5,216,045
2001	2,940,367	725,434	1,246,964	175,395	5,088,161	5,282,085
2002	2,945,334	727,415	1,260,164	174,204	5,107,118	5,348,938
2003	2,949,789	729,363	1,273,504	172,997	5,125,653	5,416,616
2004	2,953,714	731,277	1,286,984	171,775	5,143,751	5,485,126
2005	2,957,094	733,157	1,300,607	170,537	5,161,395	5,554,479

Before comparing the per capita and per customer forecast approaches, it is useful to review material on fitting trend lines and curves presented in chapter 3. The discussion considers the result of fitting linear, logarithmic, exponential, and power curves to the historic per capita water use data in Table 4-1.

The example presented in this chapter has been constructed to highlight the value of disaggregating extrapolations to the level of major customer groupings. Separate analysis of water use rates of customer groupings provides valuable information about the structure of the water system and can provide additional reliability in forecasting.

Table 4-2 Water customers and population—historic data and forecasts

Historic Data					
Year	Residential	Industrial	Commercial	Public	Population
1982	25,726	269	3,478	120	55,311
1983	26,727	289	3,466	125	56,127
1984	27,427	290	3,531	125	56,225
1985	27,928	281	3,541	126	56,694
1986	29,329	272	3,542	128	58,071
1987	30,230	272	3,483	131	58,949
1988	31,431	276	3,530	132	59,719
1989	31,932	267	3,538	134	60,351
1990	32,633	272	3,592	140	60,697
1991	32,867	277	3,554	141	63,420
1992	33,450	283	3,565	147	61,548
1993	34,134	289	3,651	150	62,124
1994	34,635	295	3,575	151	62,343
1995	34,835	301	3,657	152	62,006
Forecast of Growth in Customers and Population					
Year	Residential	Industrial	Commercial	Public	Population
1996	35,358	304	3,684	153	62,936
1997	35,888	307	3,721	154	63,880
1998	36,426	310	3,758	154	64,839
1999	36,973	313	3,795	155	65,811
2000	37,527	316	3,833	156	66,798
2001	38,090	320	3,872	157	67,800
2002	38,661	323	3,910	157	68,817
2003	39,241	326	3,949	158	69,850
2004	39,830	329	3,989	159	70,897
2005	40,427	332	4,029	160	71,961

The outlines of how to adjust water demand forecasts for anticipated changes in water prices or conservation policies are sketched in this chapter. Further details, including likely values for price and conservation effects, are provided in chapters 9 and 10.

Trending Per Capita Use Rates

The first step in projecting per capita water use over the forecast period is to obtain the historic record of total water sales and population. The per capita water use ratio over the period of historic record (shown in Table 4-3) is computed by dividing total water sales (Table 4-1) by population (Table 4-2).

Table 4-3 Water use per customer and per capita—historic data and forecasts, *thous gal*

Year	Residential	Industrial	Commercial	Public	Per Capita
Historic Data and Forecasts					
1982	104.2	3,107.4	300.9	1,337.2	85.4
1983	75.2	2,583.0	328.6	1,375.3	72.5
1984	101.2	2,343.1	314.7	1,176.6	83.8
1985	107.0	2,118.1	313.8	1,172.5	85.4
1986	80.6	3,005.3	328.3	1,609.4	78.4
1987	99.6	2,096.4	332.0	1,198.9	83.0
1988	85.1	2,336.6	320.1	1,207.1	77.2
1989	93.0	2,256.4	318.6	1,548.5	81.3
1990	77.3	2,729.1	322.6	1,353.3	76.0
1991	101.6	2,286.4	336.9	1,472.1	84.8
1992	82.4	2,397.5	311.9	1,250.5	76.8
1993	86.7	2,174.9	310.6	1,040.9	78.5
1994	77.7	2,516.8	320.4	1,066.3	76.1
1995	86.3	2,694.2	317.2	1,178.2	83.1
Water Use Coefficient Forecasts					
Year	Residential	Industrial	Commercial	Public	Per Capita
1996	82.3	2,352.1	321.1	1,185.7	78.8
1997	81.2	2,335.8	321.3	1,172.5	78.6
1998	80.2	2,319.4	321.5	1,159.4	78.4
1999	79.2	2,303.1	321.7	1,146.2	78.3
2000	78.2	2,286.7	321.9	1,133.0	78.1
2001	77.2	2,270.4	322.1	1,119.9	77.9
2002	76.2	2,254.1	322.3	1,106.8	77.7
2003	75.2	2,237.7	322.5	1,093.6	77.5
2004	74.2	2,221.4	322.6	1,080.5	77.4
2005	73.1	2,205.0	322.8	1,067.4	77.2

This per capita water sales series shows considerable year-to-year variation. Thus, Figure 4-1 graphs the annual per capita water use series from Table 4-1, together with a four-year moving average of these values.* Note that the moving average has a slight negative slope. Weather variation can be the primary contributor to the volatility of the per capita water use series. Thus, weather normalization, as discussed in chapter 6, may

*Denoting the per capita use in year t by y_t, the four-period moving average M is computed as

$$M_{t+3} = \frac{y_t + y_{t+1} + y_{t+2} + y_{t+3}}{4}$$

for $t = 1$ through 11 (since there are only 14 observations).

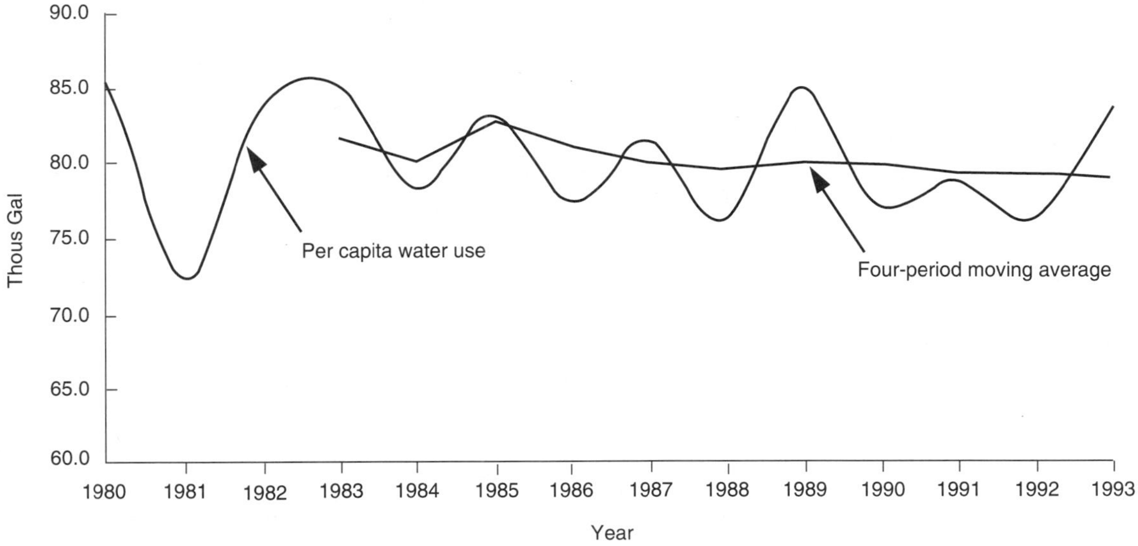

Figure 4-1 Annual per capita water use 1980–1993 and four-period moving average

be used to reduce this volatility. For purposes of presentation here, trends are analyzed with respect to the raw or unprocessed per capita water use series in Table 4-1.

A *trend* refers to a relationship between time and the change in the value of some variable, in this instance per capita water use. The simplest is a linear trend, which can be fitted by the equation

$$\text{per capita water use} = \text{constant} + \beta \times \text{time} \tag{4-1}$$

The slope or linear trend parameter β indicates a constant amount by which per capita water use will increase (or decrease) each time period.

To fit this relationship with a spreadsheet regression program requires that the year-by-year per capita water use figures be placed in one column and the time indexes, in another column. Note that any series of numbers that increase by constant amounts, such as 1,2,3,..., could be used as the time index series. The dependent variable in the regression is the column of per capita water use numbers, while the independent variable is the column with the time indexes. The regression should allow for estimation of a constant term.

There are many possible trending equations besides a straight line or linear trend. Common choices include smooth curves produced by logarithmic, exponential, and power transformations of the time variable. If the trend appears to be increasing or decreasing, but at a slower rate over time, logarithmic and exponential functions of the time variable may produce a better fit.

Goodness of fit can be assessed by computing the sum of squared errors, as described in chapter 3. Alternatively, the squared errors can be transformed into the *coefficient of determination*, or R^2. A perfect fit between the trend line and historic data is signaled by an R^2 equal to 1, while the existence of no linear relationship whatever is implied by an R^2 of 0.

The per capita water use data in Table 4-1 shows a high degree of randomness. Thus, none of the trending equations explain more than a few percent of the total variation in this series. The remedy for this might be to preprocess the historic per capita water use series, as noted earlier, but for now, trending projections based on the raw data will be used.

Figures 4-2, 4-3, 4-4, and 4-5 show linear, logarithmic, power-curve, and exponential[1] trend lines for the per capita water use data in this example. These graphs also exhibit the trending equations generating the trend line.

The values for per capita water use in the forecast 1996–2005 are generated by the logarithmic equation

$$y = -1.0908\log(x) + 82.126 \tag{4-2}$$

Where:

y = the dependent variable, per capita water use

x = the independent variable, time indexes 0 = 1980, 1 = 1981, and so forth

The negative sign on the logarithm to the base 10 of 'year' indicates that per capita water use with this trend model decreases slightly throughout the forecast period. The coefficient of determination, or R^2, indicates that approximately 4 percent of the variation in the per capita water use data is explained by the trend. Although this is the highest R^2 of the equations fit to this data, the regression in Eq 4-2 is a relatively weak forecasting relationship for this water use series. Gradually declining or linear trend lines appear safer than exponential trend lines in this water system. These trend curves are chosen here when other criteria, such as the fit of trend lines to the data, are approximately the same.

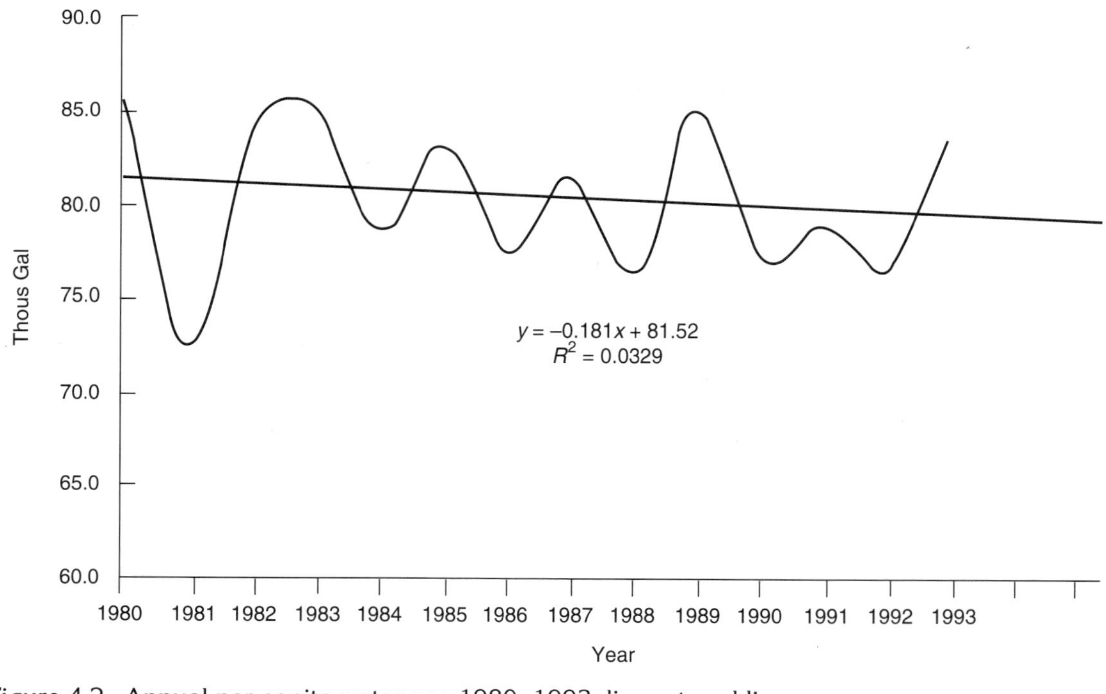

Figure 4-2 Annual per capita water use 1980–1993, linear trend line

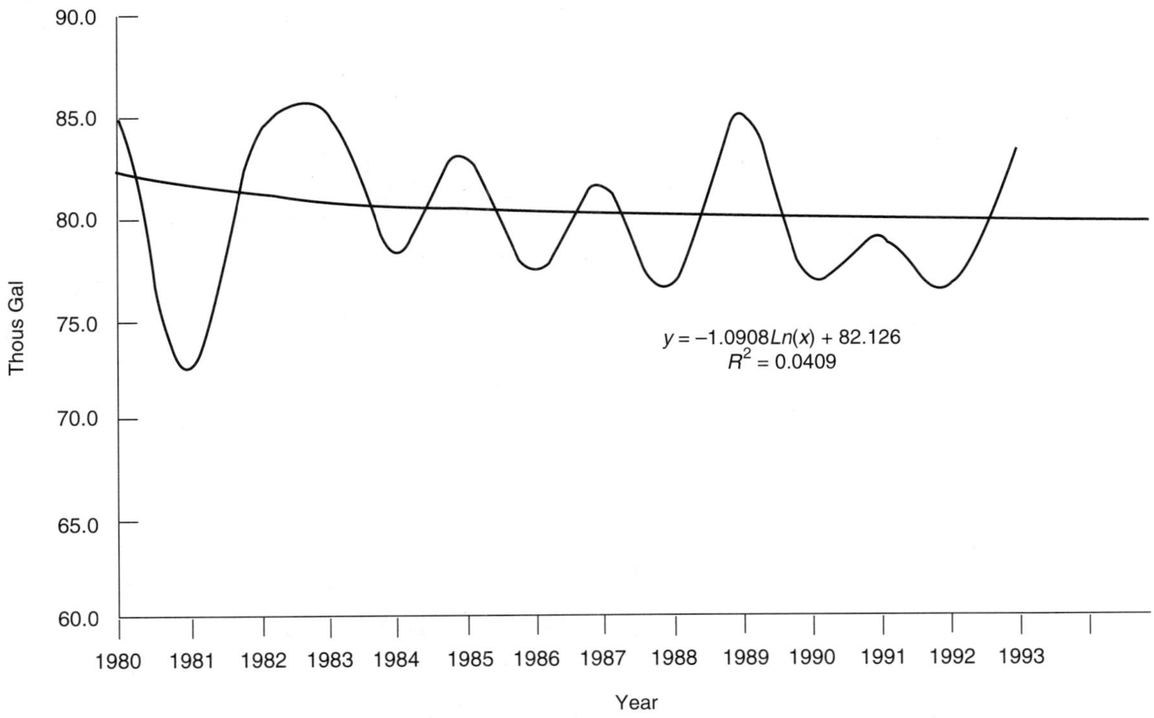

Figure 4-3 Annual per capita water use 1980–1993, logarithmic trend line

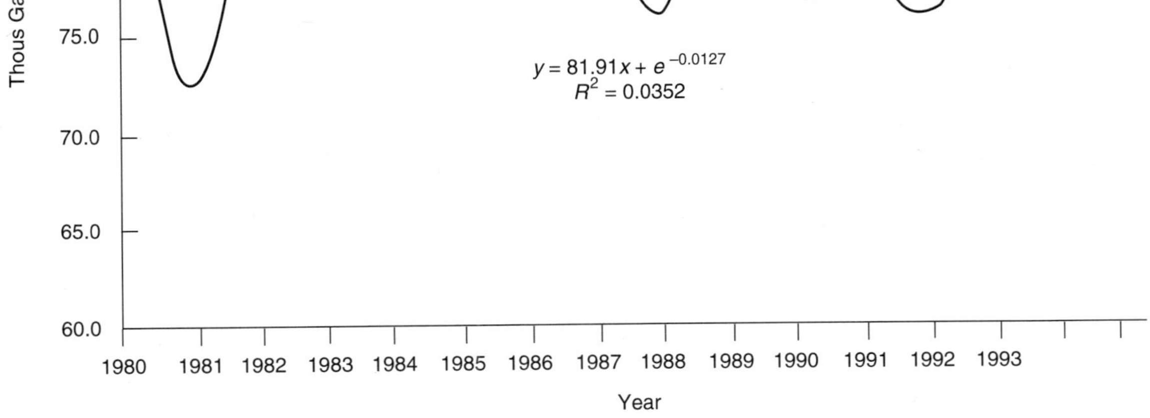

Figure 4-4 Annual per capita water use 1980–1993, power-curve trend line

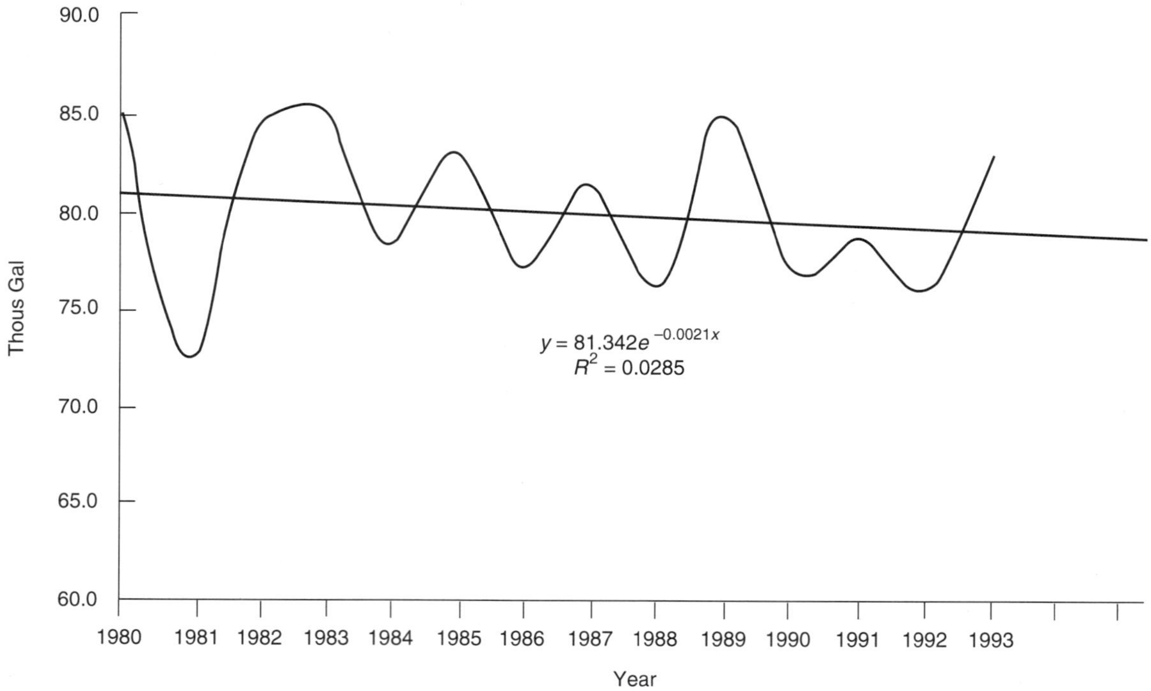

Figure 4-5 Annual per capita water use 1980–1993, exponential trend line

Using regression to establish the trend line is superior to simpler arithmetic approaches, however, because it usually assures that the designated curve will achieve a tighter fit to the data.[2]

Per Capita Forecasts

The logarithmic trending equation, associated with a declining trend that levels off, is used to forecast the 10 years of per capita water use in Table 4-3.

The population forecast is based on a growth rate of 1 percent per year, which is approximately equal to the average US population growth rate forecast for this period. Generally, local and regional demographic authorities would be the source of assumptions about future population growth.

Multiplying the estimate of future per capita water use by the population projection produces the per capita forecast of total water sales shown in the rightmost column of the bottom panel of Table 4-1.

Forecasting With Unit Water Use Coefficients

Similar procedures to those of per capita forecasts can be applied in developing forecasts of water use by customer class. The final step in the per customer forecast approach is to sum together the annual per customer forecasts to estimate total system water use over the 10-year forecast period.

Again, the first step is to examine the historic data. Several distinct trends in per customer water use are apparent in Table 4-3. Total water sales and water sales for residential, industrial, commercial, and public customers appear to be relatively static over the period of record from 1982 to 1995. Since the number of customers served in these major categories is increasing (Table 4-2), simple arithmetic suggests a decline in unit water use rates for some customer groups.

Underlying factors, such as increasing numbers of townhouses on smaller lots, could be responsible for the observed decline in residential unit water use rates. The water service area in question belongs to a larger metropolitan region and has little developable land. New residences may use less water, both owing to the fact that they generally have appliances and fixtures with higher water efficiencies and because they tend to be on smaller lots.

Industrial water use rates also show evidence of a negative trend during 1980–1993. Here, in-house recycling of water for cooling and more water-efficient processes probably account for increased efficiencies.

Persons per household drops in this water system from around three in 1982 to less than two by 1995. This trend is found in many urban areas and is often associated with an aging population as well as more single-parent households and families with fewer children.

Trending equations for residential, industrial, commercial, and public customers are estimated for linear, exponential, logarithmic, and power-curve patterns. Since trends in these raw data are extremely weak, the linear extrapolation equations are used for the sake of simplicity. The reader may refer to the data diskette for further information on the trending equations for these customer classes, and can explore the application of some of the nonlinear trending equations.

Forecasts of the number of residential, industrial, commercial, and public customers during the period 1996–2005 are based on growth rates of 1.5 percent/year for residential customers, 1 percent/year for industrial and commercial customers, and 1 percent/year for public customers. These growth rates can be assumed to be the result of studies based on projections of municipal planning authorities.

The final step in developing the per customer forecasts is to multiply the customer water use coefficients by the projected numbers of customers. This produces the 10 years of total water demand forecasts shown in Table 4-1. Total system water use here is assumed to equal the sum of estimated water use for residential, industrial, commercial, and public customers. Further disaggregation may be possible and desirable.

Comparing the Forecasts

The per customer and per capita water use forecasts parallel each other relatively closely at the beginning of the forecast period. This similarity of the per capita and per customer forecasts depends on the similarity of the growth rates for water use, customers, and people. Thus, per capita and per customer forecasts will move together if water use per capita and water use for each customer class are increasing at the same rate, and population and customer growth are occurring at the same rate. Differences in the rates of change of water use per capita and per customer and differing rates of growth of population and customers, on the other hand, can produce divergent per capita and per customer forecasts.

Divergence Between Forecasts—The Underlying Dynamic

Divergence in forecasts is illustrated in Table 4-1. Total system water demand in 2006 is predicted to rise to 5,554,479 thous gal in the per capita forecast, but to only 5,161,395 thous gal in the per customer forecast.

It is useful to study the dynamic underlying such divergences in per capita and per customer forecasts more closely. Table 4-4 provides an opportunity to do this. Table 4-4 lists annual use rates and numbers of customers for two generic customer classes over a 15-year period. Assume there are only two types of customers served by the water system in this example. Their water use coefficients are changing in different directions—one is increasing, but at a slowing rate, while the other is decreasing along a straight line. The number of customers in each customer grouping and the total population are assumed to grow at identical rates of 1 percent per year. Given 1 percent annual growth, total water use per capita first slowly increases and then gradually drops. After 12 years, the reduction in the second customer use rate in the example overwhelms the slowing rate of increase in the first water use rate, causing overall water use per capita to decrease.

Given 1 percent annual growth, total water use per capita first slowly increases and then gradually drops. After 12 years, the reduction in the second customer use rate in the example overwhelms the slowing rate of increase in the first water use rate, causing overall water use per capita to decrease.

Suppose, however, that we do not have 15 years of historic data, but are attempting to develop water demand forecasts in year 10 of this example. Then, the per capita use rates data suggest gradual increases in future per capita water use rates. However, when one calculates and inspects the use rates of the two customer groupings, the possibility for change in direction of per capita water use can be detected.

A similar situation can be created in the example of Tables 4-1, 4-2, and 4-3, when the number of industrial and commercial customers is forecast to decrease at a 1 to 2 percent rate per year. Substantial differences between the per customer and per capita forecast are thereby introduced.

Value of Per Customer Forecast Approach

The example presented suggests that dividing total water sales into residential, industrial, commercial, and public water sales can increase forecast accuracy.

Disaggregation to the level of customer groupings, therefore, can help identify and project subtrends that may be masked in the per capita approach.

Table 4-4 Use rates for generic customer classes that lead to a switch in the direction of per capita water use

Time	Use Rate of Group 1	Use Rate of Group 2	Group 1 Customers	Group 2 Customers	Customer Water Use Total	Population	Per Capita Water Use
1	100.00	99.10	80,000	60,000	13,946,000	160,000	87.2
2	104.85	98.20	80,800	60,600	14,422,964	161,600	89.3
3	107.69	97.30	81,608	61,206	14,743,733	163,216	90.3
4	109.70	96.40	82,424	61,818	15,001,517	164,848	91.0
5	111.27	95.50	83,248	62,436	15,225,374	166,497	91.4
6	112.54	94.60	84,081	63,061	15,428,181	168,162	91.7
7	113.62	93.70	84,922	63,691	15,616,776	169,843	91.9
8	114.56	92.80	85,771	64,328	15,795,220	171,542	92.1
9	115.38	91.90	86,629	64,971	15,966,122	173,257	92.2
10	116.12	91.00	87,495	65,621	16,131,254	174,990	92.2
11	116.79	90.10	88,370	66,277	16,291,874	176,740	92.2
12	117.39	89.20	89,253	66,940	16,448,910	178,507	92.1
13	117.95	88.30	90,146	67,610	16,603,059	180,292	92.1
14	118.47	87.40	91,047	68,286	16,754,864	182,095	92.0
15	118.96	86.50	91,958	68,968	16,904,752	183,916	91.9

Moreover, disaggregation of water demand forecasts offers an opportunity for increased insight into the functioning of the water system. Examining historic trends in customer water use rates can highlight possibilities for future changes in per capita demand. Considering how future use rates of customer groupings might change can highlight important policy options or the need for policy response.

Further disaggregation of water use categories is also possible. The IWR-MAIN model, version 6.0, for instance, allows for disaggregation spatially (e.g., geographical areas and pressure zones), temporally (e.g., annual, seasonal, monthly, and daily maximum), sectorally (e.g., residential, industrial, and governmental), and functionally (e.g., indoor/outdoor, landscape irrigation, cooling, and other end uses) (Dziegielewski, Strus, and Optiz 1993).

As noted in chapter 2, there can be risks in attempting to subdivide customers into too many categories. If the numbers of customers in the defined groupings fall too low, advantages of averaging over individual differences or noise can be lost. This may be one reason per capita forecast methods tend to be more prevalent in smaller water systems.

Forecast Accuracy

Fluctuations of actual water use from per capita or per customer water use trends and forecasts can be relatively large, although, over several years, positive and negative variation tends to balance out.

There are several perspectives on how to achieve greater accuracy in these forecasts, some of which are discussed in the following paragraphs.

Weather variables can contribute to a more accurate estimate of underlying trends in water use series. Thus, one might calculate annual average temperature and precipitation with long historical series. Then, annual average temperature and precipitation over the period of record for the water use data can be expressed as deviations from these long-term averages. These weather variables, expressed in deviations form, can be used as explanatory variables, along with the time trend index, in a regression that has annual per capita or per customer water use as the dependent variable. The resulting relationship will achieve a closer fit to the historic data. The resulting trending equation can be more accurate in forecasts also, because it discriminates the underlying direction in water usage from temporary high- or low-demand periods related to hot and dry or cool and wet conditions.

More accurate forecasts two or three years out in the planning horizon may be attainable through the use of fully specified water demand models. If the focus is immediate, perhaps on the next few months, improvements in forecast accuracy can be achieved by developing suitable univariate time-series models.

Some commentators have argued that unit use coefficient models have advantages in the longer term, because they require fewer forecasts of explanatory variables. A regression model with 20 explanatory factors may achieve a close fit to the historic data, for example, but each explanatory variable must be projected to use the regression for forecasting. This suggests that projections of multiple explanatory variables could introduce a greater cumulative forecast error. This rationale is sometimes given for using simple per capita water use forecasting for 10- to 20-year water demand forecasts.

A somewhat different perspective, however, is suggested by the forecasting literature. Averaging forecasts have been shown, in many contexts, to produce superior forecast accuracy (Newbold and Bos 1990). With respect to the outlook provided by Table 4-1, therefore, the question of which is the best forecast may be somewhat beside the point. The best forecast may be an average of the per capita and per customer forecasts in this application.

Adjusting Forecasts for Price and Conservation Effects

Trending is based on the assumption that the future will be similar to the past. When there are changes in water demand factors, such as scheduled rate increases or implementation of conservation programs, adjustments should be made to unit water use coefficients.

Some water resource planning and forecasting discussions draw a distinction between the unconstrained and constrained water demand forecasts. When forecasts are based exclusively on the continuation of past patterns and trends, they are called *unconstrained forecasts*. Forecasts adjusted for anticipated changes in price, conservation policies, or other relevant developments are called *constrained forecasts*.

Capacity planning provides a useful illustration of this distinction. Often, it makes sense to generate "quick and dirty" or "back-of-the-envelope" estimates of future water needs by multiplying current per capita water use by population growing at the current rate. This unconstrained forecast may suggest the need for increased structural water supply projects. The cost of these new facilities, however, can require rate increases to pay for them. If imposed, these rate increases will reduce water demand in systems where commodity charges are in effect. Factoring in these price-induced demand reductions produces a constrained forecast. Several passes at determining the relevant timing, cost, rate impacts, and estimates of the constrained forecast usually are necessary in this approach.

Elasticities provide a quick way to make adjustments for anticipated changes in prices and other demand factors, such as income. A price (or income) elasticity indicates the percentage change in water sales that is likely to result from a given percentage change in price (or income). Thus, a price elasticity of –0.3 means that, other things being equal, a 10 percent price increase will result in a 3 percent reduction in water use. This suggests that per capita water demand in year t should be reduced by 4.5 percent from the unconstrained forecast when a 15 percent increase in rates and charges is scheduled to be implemented in year t.[3] For further discussion of such adjustments, see chapter 9.

Water conservation is usually included in forecast adjustments by subtracting an estimated reduction in water use from per capita demand. This is a simple concept, but it may be challenging to frame realistic expectations about likely water savings, market penetration of various conservation devices, and the stabilty of customer habits (see chapter 10). Often, on-site statistical study (using controlled comparisons of customers with and without specific water conservation devices) or scientific sampling and regression modeling can be helpful.

Summary

This chapter presented unit water use coefficient forecasts using example data. Linear, logarithmic, exponential, and power-curve trending equations were explored. Conditions under which per capita and per customer water use forecasts for a system might diverge dramatically were noted. The enclosed data diskette contains data for the example presented in this chapter.

Disaggregating system water use by customer class promises improved accuracy of water demand forecasts by tighter explanation of component water use trends. The greatest value of disaggregation to major customer classes, however, is that developing disaggregated forecasts encourages thought and research about diverse trends that contribute to total water system sales.

The advantage of the per capita approach is its simplicity. There also are fewer variables to project in order to generate the forecast. This approach may have advantages for very long-range forecasts in which projections for almost all socioeconomic variables are linked to long-range population forecasts.

Adjusting water use coefficients for anticipated changes in price and conservation policies, as well as other predictable developments, increases the overall accuracy of the resulting forecasts. For example, it is possible to calculate how much per capita demand ought to be adjusted to take into account developments in pricing or water conservation policy (see chapters 9 and 10).

It is necessary to exercise judgment when using per capita and per customer forecasting methods. The best way to develop an informed opinion is to analyze and understand the fundamental drivers of water use by customer class in this system. This requires statistical modeling of water use against various predictor variables, such as weather, price, income, and conservation programs. These and other variables will be discussed in subsequent chapters in this book.

Notes

1 The exponential model in Figure 4-5 is an algebraic transformation of the familiar percentage adjustment approach to per capita water use forecasts. If the per capita water use increases (or decreases) according to a fixed percentage rate each year, the time pattern of per capita water use and the residuals of per capita water use from the base period average increase exponentially. This is easily seen by the fact that, if r is the annual percentage rate of increase of per capita water use, a factor of $(1 + r)^s$ is applied to the current per capita water use to arrive at the estimate of the per capita water use s years in the future.

2 Typical of the simpler arithmetic approaches is a method that subtracts the first from the last value in the historic per capita water use series and divides by the number of years between them. Better is a variation that takes the average of the first few observations and the last few observations. This calculation would allow the forecast to be built up by adding or subtracting this constant increment from the 1993 observation and continuing this process of addition or subtraction year by year. This procedure could come close to the regression trend line in some instances, but generally does not produce the curve of best fit to the historic data. There also is the problem, as seen in these data, that the historic per capita use rates jump around from year to year, making the choice of the values used to determine the annual increment problematic. Eyeballing a trend might be more accurate than using only first and last values.

3 Using price as an example, let the price elasticity of demand be e_p, the base or initial price level P_o, the new price level P_1, and the unconstrained per capita demand forecast in year t be C_t. The adjusted per capita demand is the quantity C.

$$C = C_t\left[1 + e_p\left(\frac{P_1 - P_o}{0.5\,(P_1 + P_o)}\right)\right]$$

$$\text{since} \quad e_p = \frac{(Q_1 - Q_o)}{(Q_1 + Q_o)} \times \frac{(P_1 + P_o)}{(P_1 - P_o)}$$

For example, if $e_p = -0.36$, $P_o = \$2.40$, $P_1 = \$3.00$, and $C_t = 5$ mgd, the proportional change in quantity $= -0.36\left(\frac{(3.00 - 2.40)}{0.5\,(3.00 + 2.40)}\right) = -0.08$, suggesting that the price increase will cut water sales by 8 percent. Thus, $C = 5$ mgd $(1-0.08) = 4.6$ mgd.

Note that the percent change in price is computed using the average of the old and new price. This makes the measured effect symmetrical in both directions, and provides a more accurate use of the elasticity coefficient, which conceptually is only valid for infinitesimally small changes in price.

References

Dziegielewski, B., C.A. Strus, and E.M. Opitz. 1993. Water Demand Management Planning With IWR-MAIN Version 6.0. Paper presented at CONSERV93, Las Vegas, Nev., December 1993.

Newbold, P., and T. Bos. 1990. *Introductory Business Forecasting*. Cincinnati, Ohio: South-Western Publishing Company.

Chapter 5

Causal/Structural Forecast Models

While unit use coefficient models provide an easy-to-apply forecasting technique, utility decision makers frequently can benefit from more detailed approaches. The causal/structural water use models discussed in this chapter are based on commonsense theories of cause and effect about the relationship between influences on urban water use and water sales. These models offer the opportunity to explore the effect of alternative assumptions and changing local conditions on the water use or demand forecast. Thus, they provide management with a valuable perspective on how future circumstances, including changing business conditions, climatic shifts, or new government policies, may impact the water system under various sets of assumptions.

Causal/structural models, typically estimated with multiple regression, require the development of an equation or system of equations to describe the essential aspects of how the world works. Such models depend on the ability of the analyst to determine the relationships among variables and also on being able to find or estimate future values of such independent variables as personal income, the real price of water, population, employment by major industry group, and the use of various water-related fixtures.

Additionally, there must be substantial variation in the values of each variable to obtain meaningful results. For example, if water rates have not changed during a 10-year study period, the price elasticity of demand cannot be estimated except insofar as consumers respond to the erosion of the real cost of water by overall inflation.

This chapter focuses initially on developing and interpreting a causal/structural regression model with example data contained on the data diskette. The data are a cross-sectional sample of observations on the water use and socioeconomic characteristics of individual residential customers, derived as a random sample from utility billing records. This type of data is called *microdata* at points during the discussion, distinguishing it from aggregate water use data for customer groupings.

Causal/structural models based on microdata are often found in scientific studies. Frequently, regressions fit to microdata on the behavior of individuals, households, or individual business firms have a relatively low explanatory power as measured by R^2. This might suggest the regressions have little practical application. However, these regressions can discriminate the effects of key explanatory variables with relatively high precision. Thus, regressions based on microdata tend to be looked to for unbiased estimates of response elasticities for use in policy analysis and as adjustments to be used in combination with simpler forecasting models.

In addition, regressions based on microdata often can forecast aggregate water use for relevant classes or types of customers with relatively tight confidence intervals, even though the confidence intervals for forecasts of the water use of a specific customer can be quite large. This is an example of the principle of noise mentioned in chapter 2. This chapter provides an example in which a regression forecasts water use of an individual residence with a 95 percent confidence interval of plus or minus approximately 100 percent. Forecasts of aggregate residential water for 100,000 customers based on the same regression, however, have a 95 percent confidence interval of ± 4 percent.

Strategies for boosting the explanatory power of regression models using time-series data are suggested in this chapter, and associated problems of errors in benchmark data and projections of explanatory variables are considered.[1] Discussion of these issues also is continued in chapter 6, where the estimation of weather influences within causal/structural water demand models is considered.

Estimating a Causal/Structural Model With Household Data

Developing a causal/structural model of water use per household allows a utility to answer such questions as "What if the price of water falls 15 percent, in real terms, due to inflation?" or "What if the proportion of households with water-conserving fixtures rises by 7 percent?"

Answers to such questions are available by substituting the new values of the specified independent variables into the model and recalculating the result. Hypothetical situations can be analyzed quickly. If the changes imposed on the model are not too large, the analyst can be relatively confident of these "what-if" results.

Sample Size

Suppose two urban water utilities plan a joint effort to collect data from a random sample of single-family households in order to develop a demand forecasting model. The required sample size depends to an extent on the unknown characteristics of the population and on the objective of the research. This means that, initially, sample size estimation must begin with a guess.

If the utilities wish to simply estimate and forecast average water use, the required sample size is obtained from the formula

$$n = (t^2 \sigma^2) / d^2$$

Where:

t = the t statistic showing the level of accuracy desired

σ = the population standard deviation

d = the margin of error the utility is willing to accept

For accuracy of ± 5 percent and a mean water use of 180,000 gal/year, d = 5 percent of 180,000, or 9,000 gal. If the utilities are willing to take a 1 in 20 chance of a larger error, a t of 2 can be chosen. Utility records of water consumption along with informed judgment suggest that the population standard deviation is 90,000 gal. Then, the sample size required to achieve a tolerance of ± 5 percent is

$$(2^2 \times 90{,}000^2) \;/\; 9{,}000^2 = 400$$

Half of the sample would be randomly selected from the service area of each utility. For more accuracy, a larger sample would be used.

If the utilities planned to estimate other variables, such as the proportion of the population with low-flow fixtures, a larger sample might be required. The required sample size is given by $(t^2 \times P \times Q)/d^2$, where P is the proportion of the population with the fixtures, Q is the proportion without, and d is the margin of error. The largest sample size is required if P and Q are each 50 percent. Given an acceptable error of 5 percent, the required sample size is 400. If the results of the survey were to be disaggregated based on a variable obtained from the survey, such as income level, the required sample size is much larger in order to obtain significant results for each subgroup.

Based on these calculations, a sample size of 400 single-family residents is assumed to be randomly drawn from utility billing records. Note that this sample size does not assure the regression model will produce accurate estimates—only that the cases sampled from utility records should present a representative picture of the water use of the population, i.e., the residential customers of these utilities.

The initial sample size would be increased to offset customers who did not respond to the survey. See appendix A, Direct Data Collection.

Explanatory Variables

It is imperative to have a sound conceptual basis for the variables in the model. The variables in this example, including the marginal price of water, fixed customer charges, household income, household size, lot size, the presence or absence of a swimming pool, and indoor water conserving fixtures, were chosen because they have been used in numerous published water demand models. Since the data is cross-sectional, weather factors are relevant only insofar as there is microclimatic variation in temperature and precipitation in the area sampled for the analysis. Weather stations are usually too sparsely located to permit collection of such data.

Note that the contrast between the water rates in these districts allows for an assessment of the effect of a price variable in the regression. This makes the resulting relationship a full-fledged demand function, as opposed to merely a water use relationship.

Annual household water use is the dependent variable.

The Data

For this example, a database representing two geographically contiguous water districts is synthesized from empirical data available to the authors. Residential customers in both districts have a similar socioeconomic profile, but the price of water differs in the two districts.

The variables in the sample database are

Q = annual water use per household, in thous-gal units, with a mean value of 205.94 thous gal/year

MPRICE = marginal price of water per thous gal

BILLFIXED = fixed service charge for water for the year

INCOME = household income per year, in dollars

HHSIZE = number of people in the household

LOT = lot size, in square feet

POOL = swimming pool (1 if present, 0 otherwise)

CONSERVATION = indoor water conserving fixtures (1 if present, 0 otherwise)

The water rate schedule is similar to that in Figure 9-1 (see page 113) with a fixed monthly service charge of $3.00 and a constant marginal price regardless of the quantity used. The marginal price in District 1 is $2.00/thous gal, while the marginal price of water in District 2 is $2.35/thous gal. The marginal price of water represents the cost to the consumer of purchasing additional quantities of water. Given this type of rate schedule, the cost is the same for all customers in a district.

Water, like other goods and services, follows the law of demand—less is purchased at higher prices—and thus the estimated coefficient on *MPRICE* should be negative.

Since the monthly service charge is constant, it has the same effect as a small decrease in household income. Thus the monthly service charge (*BILLFIXED*) is subtracted from income (*INCOME*) and the resulting variable *INCOME** = (*INCOME* – $36) is used in the model. Since the differences between *INCOME* and *INCOME** are so minor from a numerical standpoint, we make no additional distinction in discussion for this adjustment.

Water use per household is known to increase with household size, so the expected sign on *HHSIZE* is positive.[2] Water use is likely to increase with lot size due to summer sprinkling.

Water use is also likely to increase with the presence of a swimming pool, due to both evaporation and backwashing requirements. If a swimming pool replaces irrigated vegetation, however, its effect on water use may be quite small.

To illustrate the effect of conservation measures on water use, a composite dummy variable representing indoor water-conserving fixtures is specified in the model. Since this composite variable equals 1 when indoor water-conserving fixtures are present, the sign on its estimated coefficient is expected to be negative. In a more complete model, a variable might be included for each type of fixture individually.

Exploratory Data Analysis

Before developing the regression model, the data collected from the sample survey and from utility records should be carefully examined.

Means and standard deviations of the dependent and explanatory variables should be computed and are provided in Table 5-1.

Table 5-1 Averages and standard deviations of sample data

Variable	Average	Standard Deviation
Annual quantity of water use	205.94	83.55
Marginal price (*MPRICE*)	2.19	0.17
Household income (*INCOME*)	23.27	11.94
Household size (*HHSIZE*)	1.92	0.82
Swimming pool (*POOL*)	0.08	0.27
Water efficiency components (*CONSERVATION*)	0.36	0.48

The frequency distribution of the sample observations on water use is presented in Figure 5-1. This curve is skewed, or left-steep, with a long tail of observations at high water use levels characteristic of a lognormal distribution. This suggests a multiplicative form for the water demand model.

Correlation matrixes, showing the sign and size of the simple linear relationship between each pair of variables, are shown in Table 5-2. The top set of numbers corresponds to the correlations in the raw data, while the bottom panel of numbers indicates the correlations between the log-transformed data. The diagonal entries in each of these triangular arrays of numbers equal 1; any variable is perfectly correlated with itself.

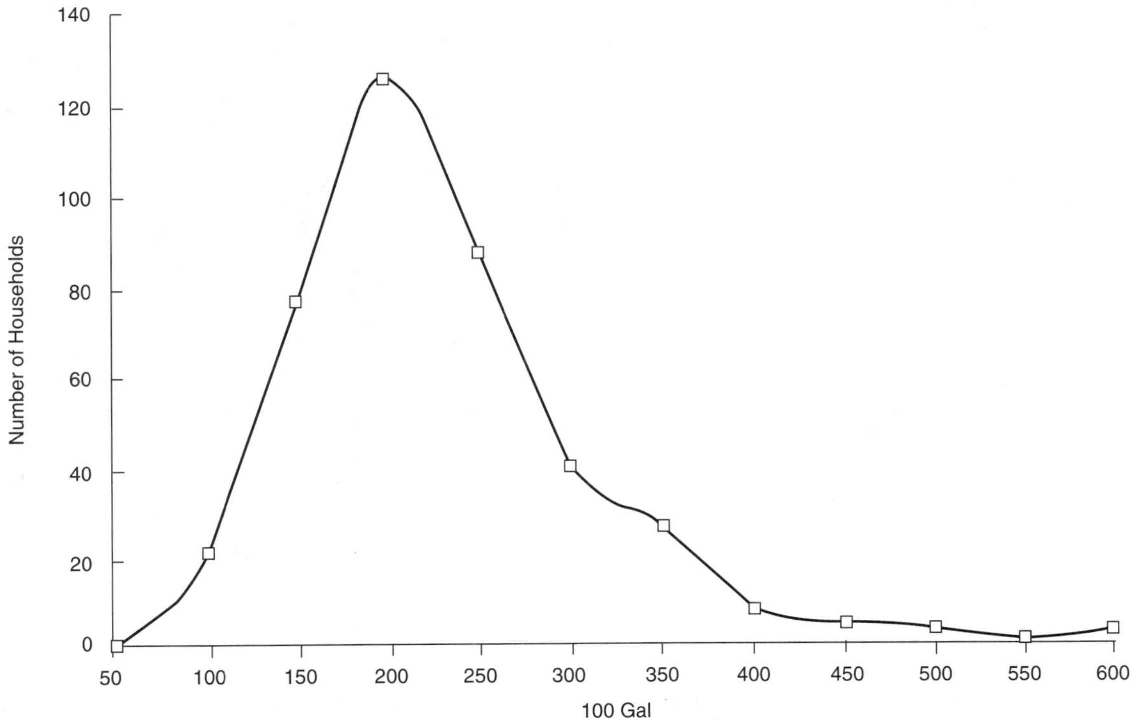

Figure 5-1 Frequency distribution of residential water use sample

Table 5-2 Correlation matrix for residential microdata—raw and log-transformed versions

	Column 1	Column 2	Column 3	Column 4	Column 5	Column 6	Column 7
	Annual quantity of water use	Marginal price	House-hold income	House-hold size	Swim-ming pool	Conser-vation	Lot size
RAW DATA							
Annual quantity of water use	1.00						
Marginal price	-0.07	1.00					
Household income	0.52	0.04	1.00				
Household size	0.17	0.07	-0.04	1.00			
Swimming pool	0.34	0.04	0.71	-0.03	1.00		
Water-efficiency components	-0.05	0.05	-0.01	0.02	-0.08	1.00	
Lot size	0.46	0.03	0.93	-0.05	0.67	0.00	1.00
LOG-TRANSFORMED DATA							
Annual quantity of water use	1.00						
Marginal price	-0.08	1.00					
Household income	0.56	0.01	1.00				
Household size	0.22	0.05	-0.02	1.00			
Swimming pool	0.31	0.04	0.53	-0.01	1.00		
Water-efficiency components	-0.07	0.05	-0.01	0.03	-0.08	1.00	
Lot size	0.47	0.00	0.87	-0.03	0.54	0.01	1.00

Several of the correlations are boosted by the log transformation. Note also the extremely high correlation between lot size and household income, as indicated in column 3 of the table. This suggests the possibility of multicollinearity, if both household income and lot size are included in the regression specification. Multicollinearity occurs when two or more independent variables in a regression are highly correlated (vary together). (For further discussion of multicollinearity, see chapter 3.) Household income has the higher simple correlation with annual quantity of water use and is the first choice for the regression specification. Thus, lot size is omitted from the model.

Regression Equation

Exploratory data analysis suggests that the effects of the explanatory variables operate in a multiplicative relationship to the dependent variable—annual water use per household. Thus, the regression model is estimated by taking natural logs of the dependent variable and each of the explanatory variables *MPRICE*, *INCOME**, *HHSIZE*, *CONSERVATION*, and *POOL*. The dummy variables *POOL* and *CONSERVATION* assume values of 1 and the base of the natural logarithms e = 2.71828, rather than 0 and 1. This is so that, after the logarithmic transformation, the values of these dummy variables will be 0 and 1, respectively.

The results of the regression can be expressed in equation form as

$$\begin{aligned} lnQ = 4.20 - 0.43ln(MPRICE) + 0.43ln(INCOME^*) \\ + 0.21ln(HHSIZE) + 0.02POOL \\ - 0.05CONSERVATION \end{aligned} \tag{5-1}$$

Where:

ln = the natural logarithm of each respective variable of the model

Other variables are as defined previously

This relationship is estimated with annual data for 400 households. The regression results and key diagnostic statistics are listed in Table 5-3.

Table 5-3 Regression analysis of sample residential water use data

Regression Statistics	
Multiple R	0.61
R^2	0.37
Adjusted R^2	0.37
Standard error	0.31
Observations	400.00

Analysis of Variance	*df*	*SS*	*MS*	*F*	Significance *F*
Regression	5.00	22.28	4.46	47.23	0.00
Residual	394.00	37.18	0.09		
Total	399.00	59.46			

	Coefficients	Standard Error	*t* Stat	*P* Value	Lower 95%	Upper 95%
CONSTANT	4.20	0.19	22.53	0.00	3.83	4.56
MPRICE	-0.43	0.19	-2.24	0.03	-0.81	-0.05
INCOME	0.43	0.04	11.75	0.00	0.35	0.50
HHSIZE	0.21	0.04	5.83	0.00	0.14	0.28
POOL	0.02	0.07	0.36	0.72	-0.11	0.16
CONSERVATION	-0.05	0.03	-1.66	0.10	-0.12	0.01

NOTE: *P* value is probability for a two-tailed test.

The R^2 of this regression indicates that the model explains 37 percent of the variation of annual water use. This is often as high an R^2 as can be expected from cross-sectional microdata, because of the noise typically generated by individual patterns of behavior. As a whole, the equation is highly significant with an F statistic of 47.23.

Significance Levels of the Estimated Coefficients

The estimated coefficients all have the expected signs. Furthermore, the coefficients on *MPRICE*, *INCOME**, and *HHSIZE* are highly significant, as indicated by their standard errors and t statistics. The *INCOME** coefficient has the highest significance level with a t statistic of 11.75.

The estimated coefficients on *POOL* and *CONSERVATION*, unfortunately, are not significantly different than 0 at the 2.5 percent level, as indicated by the inclusion of 0 in their 95 percent confidence intervals. The appropriate test of significance uses only one tail of the t distribution, since we are testing the hypothesis that the presence of a pool increases water use and conservation devices reduce it. If the hypothesis were simply that water use would change, a two-tailed test would be appropriate. With a t statistic of 1.66, *CONSERVATION* is significant at the 0.05 level, while *POOL* is only significant at 0.40. One option is to believe that the sign of the estimated coefficient is correct, but to accept the higher odds that you are in fact being misled. Thus, there is a 1 in 20 chance that *CONSERVATION* has a positive coefficient, but more than a 1 in 3 chance that *POOL* decreases water use. The other option is to omit one or both of the relatively insignificant variables and reestimate the equation without them. Since the sign of the effects of these variables is confirmed by the correlation matrix, the best decision would seem to be to leave them in the specification. On the other hand, if such variables are not highly correlated with the other explanatory variables, dropping them will have little effect on the remaining coefficients when the model is reestimated. The model can be reestimated dropping one or both of these variables using the data diskette.

Strategies for improving the statistical reliability of estimates of conservation effects are considered in chapter 10.

Coefficient Confidence Intervals

Because Eq 5-1 is in log-log form, the coefficients are elasticities. This makes impact analysis and "what ifs" particularly easy. In addition, the standard errors of *MPRICE*, *INCOME**, and *HHSIZE* are small enough (and the t statistics large enough) to provide informative confidence intervals for these elasticity or impact estimates.

As an illustration, let us derive confidence bounds for the effect of a 15 percent rate increase. Assume both utilities want to increase water rates to meet inflation in operating costs. Management is interested in exploring the probable impact on residential water use of a 15 percent rate increase.

The price elasticity of demand is estimated at –0.43, which is relatively inelastic, as expected. Its t statistic of 2.24 is significant at the 0.025 level, which means that there is less than a 1 in 40 chance that the true value is equal to or larger than 0.

Other things being equal, this price elasticity suggests that a 15 percent increase in *MPRICE* reduces residential water sales by $(-0.43) \times (0.15) \Rightarrow -6.45$ percent, while increasing utility revenue from sales to residences by 8.55 percent. Here, a handy rule is that the price elasticity of revenue equals 1 plus the price elasticity of demand (see chapter 9).

What are the confidence bounds on this impact estimate? Table 5-3 indicates that the standard error of the coefficient of *MPRICE* is 0.19. Therefore, the 95 percent confidence interval can be computed as

$$-0.43 + (1.96) \times (0.19) \geq \text{price elasticity} \geq -0.43 - (1.96) \times (0.19)$$

Thus, it is possible to assert odds of only 1 in 20 that the residential price elasticity will not be between -0.058 and -0.80.[2] (These values differ slightly from the confidence interval in Table 5-3 due to rounding error.)

The regression analysis, therefore, suggests that the impact of a 15 percent rate increase would reduce residential water demand by between 1 and 12 percent, while increasing revenues from this sector by between 14 and 3 percent, respectively.

The *INCOME** coefficient of 0.43 means that an increase in real (inflation adjusted) income per household of 10 percent should increase water sales by 4.3 percent, other things being equal. As an extension of the analysis, the reader might compute the confidence bounds for income.

Another relevant exercise is to reestimate the regression, dropping *INCOME** and including, instead, *LOTSIZE* in the specification. What would be the result of an ordinance of lot sizes in new construction, for example? How tight is the confidence interval for the estimate of this impact?

Single-Household Forecast

The regression in Eq 5-1 provides valuable structural information, but its forecasts of residential water use appear to be located in an extremely broad band. Careful analysis, however, shows that while the confidence intervals for forecasts of water use of specific residences—or a residence having specific values for the explanatory variables—are plus or minus nearly 100 percent, confidence intervals for a forecast of aggregate residential water use become tighter and tighter as more residences are included in the forecast.

The variability of forecasts from a regression depend on (1) the variance of the regression or its goodness of fit to the data, and (2) the degree to which projected values of explanatory variables of the regression deviate from their average values in the sample data.

Thus, the most reliable forecast from a regression model occurs when it is reasonable to assume that the explanatory variables will take on values equal to their sample averages. This is a type of "business as usual" prediction, in which the only source of variability is the fit of the regression to the sample data.

For the regression in Eq 5-1, we may explore this type of steady state forecast of water use of an "average residence" through the following procedure:

1. Determine the estimate of the variance of the regression. This is usually provided on the printout or spreadsheet listing the regression results.
2. Compute the square root of this variance, obtaining the standard deviation (which is itself often given on the printout as the standard error).
3. Multiply the estimated regression coefficients by associated explanatory variable values that equal the averages for the sample. Add these products to estimate average water use for the sample.

4. Build a 95 percent confidence interval around the resulting regression estimate for the average sample values.

The standard approximation to the variance of the regression estimate is the residual sum of squared errors divided by the regression degrees of freedom. In Table 5-3, this number is in the middle panel in the line titled "Residual" and column titled "MS."

The square root of the variance is the standard deviation of the regression error process. Thus, $\sqrt{0.094372}$ is 0.3072, which is shown as the standard error in the top panel.

Remember that this standard deviation (error) refers to the log-transformed data, thus the following discussion refers to averages of the log-transformed values of explanatory variables. The sample average of the log-transformed water use numbers in the dataset is 5.253 = $ln(191.2)$. Substituting the average values of the log-transformed explanatory variables in the regression Eq 5-1 produces approximately this same number.

Suppose that next year we anticipate that the average log-transformed values for *MPRICE*, *INCOME**, *HHSIZE*, *POOL*, and *CONSERVE* will precisely equal their averages in this sample data. Because there is considerable unexplained variation in this sample data, it is possible that the average of the log-transformed water sales figures for next year will be quite different than 5.253.

Given 400 observations, a 95 percent confidence interval is associated with a t value of 1.96. Thus, the confidence bounds are $5.253 - 1.96 \times 0.3072 = 4.651 = ln(104.7)$ and $5.253 + 1.96 \times 0.3072 = 5.855 = ln(349)$. Converting these numbers into 1,000 gal, we ascertain that a 95 percent confidence interval for the regression estimate of the average log-transformed water use in the sample is between 104.7 and 349 thous gal. These 95 percentage confidence bounds are –55 to +83 percent of the predicted point estimate of 191 thous gal. Thus, while the regression rather precisely estimates effects for certain of the explanatory variables, it does not forecast water use of an average household with much accuracy at all.

Some analysts prefer linear rather than double logarithmic demand models. When this same model is estimated in strictly linear form, average water use is 180.3 thous gal/year and the standard error of the regression is 68.9 thous gal/year. This yields a 95 percent confidence interval of

$$180.3 \pm 68.8 \times 1.96, \quad \text{or} \quad 180.3 \pm 135,$$

which is similar to that found from the double logarithmic model.

Total Population Forecast

The confidence interval for a forecast for the combined water use of all of the utility's customers is dramatically smaller. For computational simplicity, this example is developed for a strictly linear regression.

For a population of 100,000, the forecast of aggregate water use when each of the independent variables of the regression equals its mean value from the sample is $100{,}000 \times 180.3$ thous = 18,030 mil gal. The variance of this forecast[3] is estimated by

$$\text{Var} = K^2 \frac{S^2}{N} + K\,S^2 \tag{5-2}$$

$$= (100{,}000)2 \times \frac{(68.9)^2}{400} + 100{,}000 \times (68.9)^2$$

Where:

K = the forecast population = 100,000

S = the standard error of the regression = 68.9 thous gal, which is taken to be the best estimate of σ, the population standard error

N = the number of observations used in the regression = 400

Thus, the standard error of the forecasts, which is equal to the square root of the forecast variance, is 345.188 mil gal. This yields a forecast confidence interval of 18,030 $\pm$ 345.2 $\times$ 1.96, or 18,030 $\pm$ 677 mil gal. Therefore, the predicted water use falls between 17,353 and 18,707 mil gal, for a forecast range of $\pm$ 3.7 percent.

If one of the independent variables takes on a value different than its mean value in the sample, the forecast variance includes an additional term that can result in a much wider interval. For example, suppose marginal water rates increase by 15 percent from their mean value (P of \$2.19 to P^* of \$2.52). Using the price elasticity from the log model of -0.43 yields a drop in water sales of -0.43 $\times$ 15 percent = 6.45 percent. The new forecast of water sales per household is 168.7 thous gal, so that aggregate use of 100,000 customers is 16,867 mil gal. The forecast variance now includes the following additional term:

$$\left[\sum_{i=1}^{k}(P^* - \bar{P})\right]^2 \times \frac{S^2}{\sum_{i=1}^{N}(P_i - \bar{P})^2} = (33{,}000)^2 \times \frac{(68.9)^2}{11.53} \tag{5-3}$$

Adding this to the previous forecast variance yields the new forecast variance, S^2 = 56,752,545 $\times$ 10^4, and the standard error of the forecast S = 753 mil gal. The 95 percent forecast confidence interval now becomes 16,867 $\pm$ 753 $\times$ 1.96, or 16,867 $\pm$ 1,477 mil gal, which is a forecast interval of 15,390 to 18,344 mil gal, or a range of $\pm$ 8.8 percent.

If more than one independent variable takes on a value that differs from its mean, the computation of the forecast variance becomes much more complex unless the correlation among the independent variables is 0. It is also difficult to estimate the confidence interval for an aggregate forecast from a model with a nonlinear dependent variable, such as the double log model, although the same principle operates.

Estimating a Causal/Structural Model With Time-Series Data

Many utility analysts use time-series rather than cross-section data for developing forecasting models. Eq 3-2 on page 27 provides an example of a demand model to be estimated with time-series data that could be used for forecasting purposes.

$$Q = \beta_0 + \beta_1\, RAIN + \beta_2\, TEMP + \beta_3\, PRICE + \beta_4\, INCOME + \beta_5\, HHSIZE + \varepsilon$$

Where:

Q = water use for each month, in thous-gal units

β_o = the constant or intercept of the model

β_i, i = 1,2,3,4,5 = estimated slope coefficients

$RAIN$ = monthly rainfall, in inches

$TEMP$ = average monthly temperature, in °F

$PRICE$ = the price per thous gal of water, in dollars

$INCOME$ = average annual income per household, in dollars

$HHSIZE$ = number of persons in household

ε = the residual error of the estimate

This model is formulated to estimate the water use of a typical household. Thus, a forecast of total residential water use requires multiplying the output of this model by a projection of the number of households. Since the data used to estimate this model consist of average values for one household, the accuracy of any forecast using this data depends on this model having a very low standard error of the estimate, or somewhat equivalently, a very high R^2. The variance for a forecast of average household water use is estimated by

$$\text{Var} = S^2\left[1 + \frac{1}{N} + \frac{(X^* - \bar{X})^2}{\sum_{i=1}^{N}(X_i - \bar{X})^2}\right] \tag{5-5}$$

Where:

S^2 = the variance of the regression estimate

N = the number of observations used in the regression

$\bar{X}$ = the mean value of X over the data set

X^* = the forecasted value of that independent variable

X_i, i = 1, 2, 3 = observed values of an independent variable

The standard error of the forecast S used in developing the confidence interval is the square root of this variance.

When all of the independent variables of the regression are forecasted to assume their average values, the third term in the equation is 0. For example, suppose in the regression S^2 = 0.57 thous gal, N = 72, and Q = 13.42 thous gal/month. The forecast variance = $(0.57)^2 \times (1 + 1/72)$ = 0.3294 and the standard error of the forecast is = 0.574.

For a 95 percent confidence interval, the t statistic for a regression with 72 observations and 6 estimated coefficients (66 degrees of freedom) is 2.00. Thus the confidence interval is $13.42 \pm 0.574 \times 2.00$, or 13.42 ± 1.15, giving a forecast interval from 12.27 to 14.57, or $\pm$ 8.6 percent. Since the forecast for the entire residential sector is this forecast multiplied by the total population, it would have the same relative confidence interval, $\pm$ 8.6 percent, assuming a perfect population forecast.

If one of the independent variables was forecasted with a value different than its mean value, the forecast interval would be wider. Eq 5-5 shows that the forecast variance increases as the predicted value of an independent variable moves away from its mean value within the sample. When more than one independent variable is forecasted with a value that differs from its mean value, the computation of the forecast interval is best carried out with matrix algebra, since it involves both variances and covariances of the independent variables.

Improving Forecast Accuracy With Disaggregation

Although the forecast confidence intervals developed in this chapter for both cross-section and time-series data were smaller than $\pm$ 10 percent, more accurate forecasts would be highly desirable. A reasonable goal might be to achieve ex post forecast model errors that are due more to unpredictable variations of the weather than to all other sources.

An excellent strategy for improving forecast precision is disaggregation of utility customers by categories, such as single-family residential, commercial, industrial, etc. For a large utility, it may be worthwhile to disaggregate customers into numerous categories in order to develop a forecasting model for each. Geographic disaggregation is also a possibility when distinctly different areas are served by the same utility.

The primary benefit of disaggregation is that accuracy can be improved by developing a different model specification for each category to reflect its individual characteristics. Disaggregation also permits the analyst to use entirely different forecast methods on various categories. For example, water use by one category of customers could be forecast with simple extrapolation, another with water-use coefficients, a third with a causal/structural model, and a fourth category with a Delphi process. (See appendix A for a brief introduction to Delphi.) Disaggregation also allows the analyst to focus attention on whichever category forecast most needs improvement.

Another benefit of disaggregation is that customer response to changes in some of the independent variables included in a regression model may differ markedly among various groups. For example, Agthe and Billings (1987) found that the marginal price elasticity of demand varied systematically by income group in Tucson, Ariz., from -0.40 to -0.57, with the lowest elasticity among the highest income group. Ignoring differences among groups of customers such as these results in a larger standard error, lower R^2, and hence less precise forecasts. Differences such as this can be accommodated by running individual regressions for each group. An often used alternative is to include intercept and/or slope dummy variables in the model to capture these effects.

Slope and Intercept Dummy Variables

An *intercept dummy variable* takes on values of 0 or 1. A change in the dummy variable from 0 to 1 causes a parallel shift in the regression line by the amount of the estimated coefficient. This is equivalent to adding the coefficient on the dummy variable to the equation constant when the dummy variable equals 1. This type of variable is often used to represent the presence or absence of a condition or status, such as the presence of a swimming pool or a low-flush toilet.

In contrast, a *slope dummy variable* is used to allow the slope or estimated coefficient of an ordinary variable to differ between two groups of customers or two situations. For example, the impact of precipitation on water use could be divided between months with a low average temperature and those with a high monthly temperature. This would require using precipitation (R) as one independent variable, and a second variable (HR) that would be equal to R during periods of high temperature and equal to 0 otherwise. Thus, the estimated effect of precipitation during periods of high temperature is equal to the sum of the two estimated coefficients. An equivalent result would be obtained by defining R as precipitation during cool months and 0 otherwise, and HR as precipitation during hot months and 0 otherwise. Variables such as this can be designed to meet the peculiarities of the utility's customers.

Cross-Section, Time-Series Models

The ideal causal/structural model probably is one using pooled cross-section, time-series data. A model using this approach is described by Weber (1989) for Oakland, Calif. Using dummy variable techniques, customers with widely varying income levels, geographic locations, water use habits, etc., can be combined into a single regression. The benefit of this tactic is a greatly increased sample size and an enhanced ability to generate precise parameter estimates. The increase in precision arises from the greater variation in the values of the independent variables and the larger sample size. For example, a data set containing wide variation in water price leads to a much more accurate estimate of the price elasticity of demand than a data set containing only a narrow range of prices. At the extreme, when there is no variation in the price of water within the data set, the price coefficient cannot be estimated.

In Eq 5-2, note that the largest component of the forecast variance shrinks in proportion to sample size; in Eq 5-3, it shrinks both as sample size increases and as the variance of the independent variable increases. Thus, even if a combined cross-section, time-series model has a higher standard error, these other effects of combined samples are likely to dominate, leading to more precise forecasts (i.e., narrower forecast intervals).

Forecast Evaluation

Forecasts should be tested against reality. The ultimate test comes when the actual values of the variables in question are available.

Differences between the actual and forecast results do not necessarily mean the forecast is poor. Large forecast errors can reflect inherent randomness and volatility in the dependent variable. The existence of such unpredictability is signaled by a wide forecast confidence interval.

There also are surprises. A forecast based on a structural regression equation attempts to integrate and project the best current guess about the future values of the explanatory variables. The demand for urban water in an area, for example, may be linked to the completion of a major infrastructure project, perhaps a port, an airport, or a highway.

The emphasis in forecast evaluation, therefore, should be on accounting for the forecast errors. Would the forecast model have predicted with greater accuracy if an emergent development or surprise did not happen? Did the error arise from the estimated model itself, from a misspecification of the model due to the omission of some variable that turned out to be important, or from errors in the forecast values of the independent variables? Identifying the source of the forecast error allows the analyst to evaluate the potential for forecast improvement, and determine how efforts toward revision should be directed. This analysis will also reveal the extent to which the analyst might have reduced the forecast error by a more thorough initial effort.

Instead of waiting for time to pass before evaluating the forecast, a backcast can be developed. This requires forecasting the variables for the time period before the beginning of the data series used in developing the model or for a "holdout" sample. The backcast results then can be compared with the actually known but unused values of the dependent variable. This comparison provides a reality check on model accuracy.

Summary

Causal/structural models use a wide range of information to develop estimates of the impact of water demand influences in order to generate forecasts.

This chapter focused on a regression model estimated with household microdata from a random sample of utility billing records and customer surveys. The discussion considered sample size issues and exploratory data analysis. The regression was estimated in log-log form, so that the estimated coefficients could be read as elasticities, suggesting the percentage change in water use that will result from a given percentage change in each explanatory variable. A confidence interval for the price elasticity was computed and used to estimate a confidence interval for the change in revenue based on a 15 percent increase in water rates.

The same model was estimated in linear form, and a 95 percent confidence interval was estimated for a forecast of the utility's total residential water sales. A second forecast was computed for a 15 percent rate hike, along with its forecast confidence interval.

A forecast confidence interval was then computed for a hypothetical time-series forecasting model using average household water sales as the dependent variable. Even though this model had an estimated R^2 of 0.97, compared with the cross-section model R^2 of 0.37, the forecast interval for the time-series model was more than twice as large!

This example, along with examination of the forecast interval formulas, suggests that forecasts based on cross-section models of individual customer water use may be more precise than those developed with more aggregated time-series data. The ideal forecasting model, however, is likely to be a combined time-series, cross section, since the data in such a model incorporates increased variability of the independent variables, leading to more precise parameter estimates, along with a larger sample size.

Notes

1 There are pitfalls in a straightforward "maximum R^2" criterion, especially when models are built with time-series data. The classic pitfall of a casual/structural model built with time-series data is that variables that increase with time can show up with spurious correlations. Regression models built with time-series data, as a result, often show much higher R^2 than regression models built with cross-sectional data.

This apparently greater explanatory power can be illusory, however, and suggest incorrect forecasts if hidden factors relating to the growth of the explanatory variable, but not related to the growth of the dependent variable, change. One precaution that may be taken is to take first differences of variables exhibiting trends. It is less likely that the time differences in two variables will show strong correlation when there is not true causal or functional interdependence than it is for their raw series to show correlation.

Another proviso pertains to univariate time-series models or to elaborate noncausal adjustments to fit or adjust the residuals in causal/structural models. Such models can contain mathematical features that can "overfit" to the sample data. Univariate time-series models that follow every contour of the sample data may not generalize or forecast as well as other models with lower R^2.

2 There are efficiencies in water use per resident on a household basis. This is, in part, because lot size and outdoor residential water use usually increase less than proportionally to the number of household residents. Thus, a single three-member household generally uses less water than three one-member households.

3 This point is sufficiently important to justify an exposition of the underlying logic. We will do this for a simple situation in which a regression is developed on a single explanatory X. The assumption is that the true relationship between a dependent variable Q and explanatory variable X is given by the relationship

$$Q = a + bX + \varepsilon \tag{5-6}$$

Where:

ε = a random disturbance or error term distributed according to a normal distribution with 0 mean and standard deviation S

According to Eq 5-6, therefore, Q is a random variable, or, more precisely, a variable with deterministic and random components.

Assume now that the classic experimental situation pertains, so that we can determine N exact values of X: $X_1, X_2, X_3, \ldots, X_N$ so as to observe the corresponding values of Q: $Q_1, Q_2, Q_3, \ldots, Q_N$.

A regression can be estimated with these N datapoints on (X, Q), as

$$\hat{Q} = \hat{a} + \hat{b}X \tag{5-7}$$

Where:

$\hat{a}$ and $\hat{b}$ = ordinary least-square regression estimates of the parameters a and b

$\hat{Q}$ = the estimate of the value of the dependent variable, when the explanatory variable assumes the value X. Now, note that

$$\overline{Q} = \hat{a} + \hat{b}\overline{X} \tag{5-8}$$

so that we may rewrite Eq 5-8 as

$$\hat{Q} = \overline{Q} + \hat{b}\,(X - \overline{X}) \tag{5-9}$$

for general X.

Suppose then that we wish to use this relationship to predict or forecast the value of Q, when $X = X^*$, and inquire into the confidence bounds for the resulting forecast. This requires identifying the variance of the forecast.

The variance operator is represented as var(.), and is defined for a discrete series of N observations on a variable X as

$$\text{var}\,(X) = \left(\frac{1}{N}\right)\sum_{i=1}^{N}(X_i - \overline{X})^2 \tag{5-10}$$

Then, we can derive an expression for the variance of the forecast errors

$$\text{var}\,(\hat{Q} - Q) = var\,[\overline{Q} + \hat{b}\,(X^* - \overline{X}) - Q] = var\,(\overline{Q}) + (X^* - \overline{X})^2\, var\,(\hat{b}) + var\,(Q) \tag{5-11}$$

Here, we have been able to apply the variance operator to the components of the original expression at the far left of the set of equations because these components are stochastically independent of each other. We also made use of the rule that var $(cX) = c^2$ var (X), when c is a constant and X is a random variable.

Each of the components in the far right expression in this chain of equalities has a well-established variance. Thus, the variance of the arithmetic average of the values of Q resulting from observation of N values, essentially a sampling of the random component of Q n times, is $\frac{\sigma^2}{N}$. This is simply to say that the variance of a sample of size N from a normally distributed variable is the variance of the population divided by N.

The variance of Q also is easy, since, by definition, this equals σ^2.

Finally, the variance of the regression estimate $\hat{b}$ is

$$\frac{\sigma^2}{\sum_{i=1}^{N}(X_i - \overline{X})^2} \tag{5-12}$$

The variance of the forecast from the actually observed value, then, is

$$\sigma^2\left[1+\frac{1}{N}+\left(\frac{(X^*-\bar{X})^2}{\sum_{i=1}^{N}(X_i-\bar{X})^2}\right)\right] \tag{5-13}$$

This expression is presented in many forecasting books, such as Pindyck and Rubinfeld (1991). Thus, if X^* equals the average value of X in the sample of N experimentally generated observations and N is large, the variance of the forecast from the actually observed value is essentially equal to the variance of the regression.

Now, let us consider the situation in which the regression in Eq 5-7 is used to predict an aggregate effect for K values of X. We are interested, then, in predicting the summation

$$\sum_{j=1}^{K} Q_i = \sum_{j=1}^{K}\left[\bar{Q}+\hat{b}(X_j^*-\bar{X})\right] \tag{5-14}$$

This is precisely the situation in which we are interested in predicting the market response as the summation of the predicted individual demands for consumers.

The relevant comparison is with the summation of the actual values for these K customers in this future time period, so we are concerned to analyze the following expression:

$$\text{Var}\sum_{j=1}^{K}(\hat{Q}_j-Q_i) = Var\left(\sum_{j=1}^{K}\left[\bar{Q}+\hat{b}(X_j^*-\hat{X})\right]-\sum_{j=1}^{K}Q_j\right) \tag{5-15}$$

This can be broken into two parts, as

$$Var\left(\sum_{j=1}^{K}\left[\bar{Q}+\hat{b}(X_j^*-\bar{X})\right]\right)+var\left(\sum_{j=i}^{K}Q_j\right) \tag{5-16}$$

since the actual demands for the K customers are independent of the predicted demands. The next steps are

$$K^2\frac{\sigma^2}{N}+\left[\sum_{j=1}^{K}(X_j^*-\bar{X})\right]^2 Var(\hat{b})+K\sigma^2=\sigma^2\left[\frac{K^2}{N}+\frac{\left[\sum_{j=1}^{K}(X_j^*-\bar{X})\right]^2}{\sum_{i=1}^{N}(X_i-\bar{X})^2}+K\right] \tag{5-17}$$

The important thing to notice about this expression is that the variance of the aggregate forecast based on the regression on microdata can be reduced by increasing the sample size N (reducing the first and second terms), by increasing the variance of the independent variable (increasing $\sum_{i=1}^{N}(X_i-\bar{X})^2$), and by reducing the variance of the overall regression.

References

Agthe, D.E., and R.B. Billings. 1987. Increasing Block Rates for Water: Equity, Price Elasticity and Household Income. *The American Journal of Economics and Sociology*, 46(3).

Pindyck, R.S., and D.L. Rubinfeld. 1991. *Econometric Models and Economic Forecasts*. 3rd ed. New York: McGraw-Hill.

Weber, J.A. 1989. Forecasting Demand and Measuring Price Elasticity. *Journal AWWA*, 81(5).

Chapter 6

Forecasting Seasonal and Peak Water Demands

Seasonal demand has important design implications for water systems, because capacity must be sized to meet increased customer purchases during peak periods. Identifying and forecasting seasonal patterns of water use also can affect budget decisions and short-term revenue forecasting for urban water utilities.

The seasonal pattern of water use and variously defined peak usage levels (e.g., hourly, daily, and monthly) typically are "moving targets" in a system that, year by year, is growing and changing. This chapter considers the following three approaches to extracting the underlying monthly pattern of water use in this context:

- averaging models
- extrapolation models, such as exponential smoothing
- regression models

These approaches have various strengths in addressing the seasonal and peak demand forecast problem.

Averaging can be recommended because of its computational simplicity. Also, it can "net out" trends in water use due, for example, to population growth, so long as patterns of the variation in customer usage are relatively stable.

Given more shifting usage patterns year by year, exponential smoothing offers valuable tools. Exponential smoothing is an adaptive modeling approach that allows for time-varying trends and patterns based on recent behavior of a time series. A Winter's exponential smoothing model is developed in this chapter to illustrate a method for simultaneously identifying the seasonal pattern and the year-by-year trend in system water demand.

These univariate approaches to forecasting seasonal and peak demands have a shortcoming. They do not explicitly include weather as an explanatory factor. Accordingly, a regression approach is discussed that incorporates temperature and precipitation variables and dummy (0/1) variables. These regression models are used to estimate the effect of hot and dry versus cool and wet conditions on system water usage.

Averaging Procedures

The averaging approach to identifying seasonal water use is straightforward and relatively robust as an estimating technique. For data, one can use a quarterly or, preferably, monthly time series on system or customer class water use. The procedure produces reasonable estimates of the time pattern of water use if weather patterns are not too volatile and other important influences, such as water commodity charges, are relatively constant or slowly changing over the historic and forecast period.

The aggregate residential water use series graphed in Figure 6-1 provides a good example. These are monthly observations showing a strong seasonal pattern with a slight upward trend at the end of the period, which extends from the beginning of 1980 through 1990.

A Simple Averaging Technique

The averaging procedure using the ratio of each month to the calendar year total is summarized in Table 6-1. The top panel in Table 6-1 lists monthly residential water sales during the period 1980–1990. The monthly water sales are also totaled by year. These annual

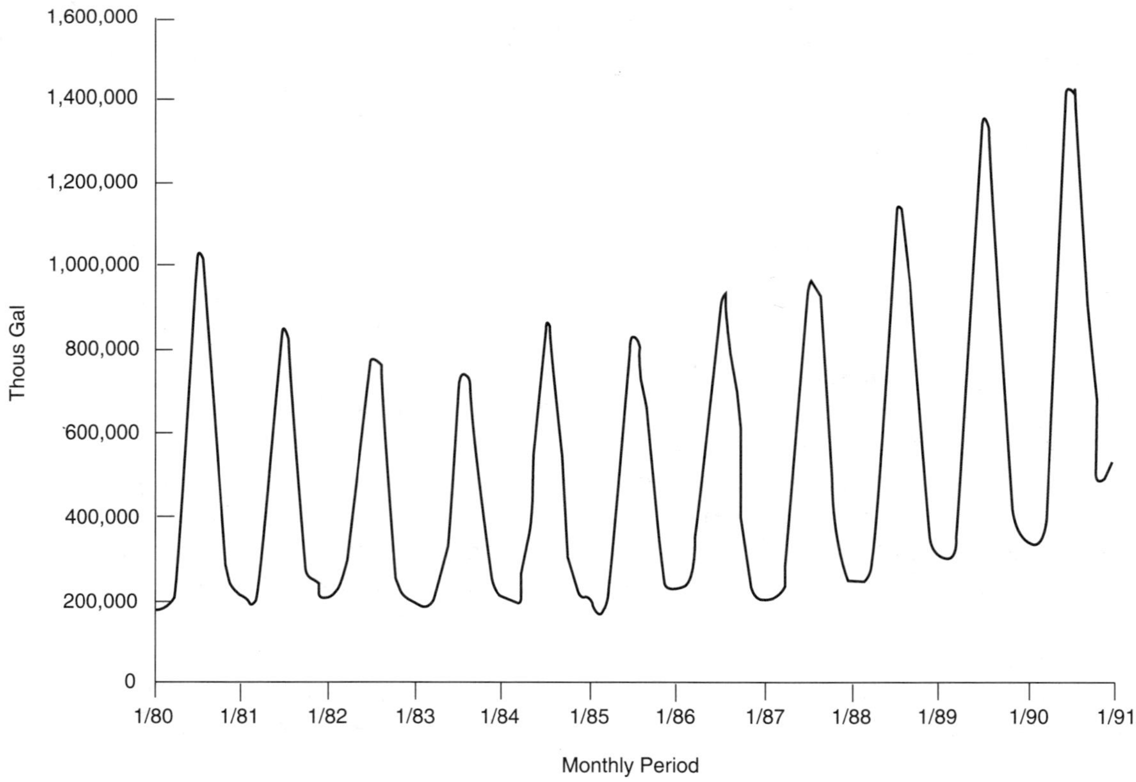

Figure 6-1 Monthly residential water use series

Table 6-1 Method of averages for identification of seasonal pattern of water use

	Monthly Residential Water Sales, in *thous gal*										
	1980	1981	1982	1983	1984	1985	1986	1987	1988	1989	1990
Month 1	181,614	216,905	210,067	196,188	217,436	212,896	225,367	219,230	259,540	316,995	360,404
2	179,091	199,175	225,344	191,145	204,262	165,494	254,723	214,585	253,424	316,831	350,829
3	188,428	219,918	273,468	202,737	209,746	203,400	303,961	237,424	282,658	368,103	370,900
4	319,818	366,069	438,504	264,172	350,891	422,868	466,971	390,495	520,851	606,494	662,336
5	534,441	506,897	537,514	347,684	525,284	584,001	589,392	539,927	778,006	817,138	1,051,602
6	909,798	747,958	693,886	577,082	790,560	814,326	828,005	839,708	1,073,232	1,174,552	1,432,110
7	1,042,919	854,482	790,127	721,065	885,728	829,413	948,082	986,998	1,167,703	1,379,976	1,440,184
8	874,878	620,157	699,638	746,667	733,198	711,973	820,001	865,472	1,077,644	1,246,329	1,209,070
9	685,648	360,670	543,905	661,886	557,794	614,755	654,965	698,297	911,392	1,006,339	966,787
10	416,671	255,707	312,978	409,134	333,000	357,435	378,665	439,375	584,657	644,723	643,323
11	267,402	251,009	222,833	275,917	239,791	251,213	252,671	315,521	425,051	463,092	495,229
12	223,461	208,830	208,802	230,022	221,168	226,467	227,054	271,769	343,464	384,139	521,907
TOTAL	5,824,167	4,807,777	5,157,055	4,823,699	5,268,858	5,394,242	5,949,855	6,018,801	7,677,619	8,724,708	9,524,680

	Ratios of Monthly to Total Annual Water Sales											Average for Month	Standard Deviation of Average of Monthly Ratios
Month 1	0.031	0.045	0.041	0.041	0.041	0.039	0.038	0.036	0.034	0.036	0.038	0.038	0.004
2	0.031	0.041	0.044	0.040	0.039	0.031	0.043	0.038	0.033	0.036	0.037	0.037	0.005
3	0.032	0.046	0.053	0.042	0.040	0.038	0.051	0.039	0.037	0.042	0.039	0.042	0.006
4	0.055	0.076	0.085	0.055	0.067	0.078	0.078	0.065	0.068	0.070	0.070	0.070	0.010
5	0.092	0.105	0.104	0.072	0.100	0.108	0.099	0.090	0.101	0.094	0.110	0.098	0.011
6	0.156	0.156	0.135	0.120	0.150	0.151	0.139	0.140	0.140	0.135	0.150	0.143	0.011
7	0.179	0.178	0.153	0.149	0.168	0.154	0.159	0.164	0.152	0.158	0.151	0.161	0.010
8	0.150	0.129	0.136	0.155	0.139	0.132	0.138	0.144	0.140	0.143	0.127	0.139	0.008
9	0.118	0.075	0.105	0.137	0.106	0.114	0.110	0.116	0.119	0.115	0.104	0.111	0.015
10	0.072	0.053	0.061	0.085	0.063	0.066	0.064	0.073	0.076	0.074	0.068	0.069	0.009
11	0.046	0.052	0.043	0.057	0.046	0.047	0.042	0.052	0.055	0.053	0.052	0.050	0.005
12	0.038	0.043	0.040	0.048	0.042	0.042	0.038	0.045	0.045	0.044	0.055	0.044	0.005

water use totals become the denominator of the ratios between monthly and total annual water use for each month and year presented in the lower panel of the table.

Averaging these ratios across the rows produces the final estimate of the average seasonal water use pattern. The method provides an estimate of the typical proportion of system water usage occurring in any month of the year. These estimates are listed in the column labeled "Average for Month" in the lower panel. Standard deviations of these averages are listed in the adjacent column at the far right.

This computation indicates that July is the peak demand month and that January is the month when water sales are lowest. Figure 6-2 graphs these ratios with an approximate 95 percent confidence interval, indicated by dotted lines.

The standard deviation of each month's ratio provides a measure of variability. Assuming that the proportion of water use occurring in a given month is normally distributed, the ratio would be expected to be within one standard deviation of its average value 68 percent of the time and within two standard deviations 95 percent of the time. Given that July's water use ratio is 0.161 of the annual total, with a standard deviation of 0.010, we would expect July water use to range from 0.141 to 0.181 of the annual total in about 19 years out of 20. This range for the proportion of monthly water use occurring in July could be combined with maximum estimates for annual water sales to identify maximum likely water needs during the month of peak water demand.

Since July accounts for a higher than average percentage of total annual water use, another month will have a lower than average water use ratio. Thus, the size of the confidence interval for each month should be interpreted only with respect to its individual

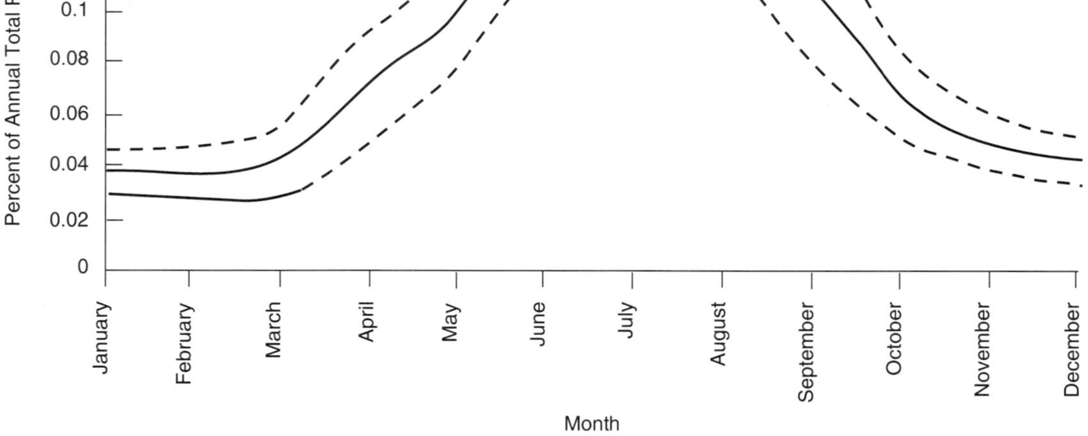

Figure 6-2 Monthly pattern of water use in the 1980s

mean values. The standard deviation for each month divided by that month's mean ratio provides a relative measure of the variability of each month's sales.

13-Month, Moving-Average Method

The widely used moving-average model is similar to the month-to-annual ratio model, except that the monthly ratios are computed using a moving average centered on each month. This model is also available in most forecasting statistical packages. The model requires the following steps, which can be easily set up on a spreadsheet.

Step One. Take a 12-month moving average of individual monthly values. Assuming the data starts with January 1980, the first entry would be placed in the row for June 1980, as shown in Table 6-2. No entries are computed for the rows for January through May 1980. If the data ends with December 1990, the last entry is placed in the row for June 1990.

Table 6-2 13-month, moving-average, seasonal model

	Date	Sales (thous gal)	12-Month Moving Average	2-Month Moving Average	Raw Seasonal Ratio	Average Seasonal Ratios	Adjusted Seasonal Ratios	Deseasonalized Sales Data
1980	Jan	181,614				0.476	0.475	382,625
	Feb	179,091				0.462	0.460	389,022
	Mar	188,428				0.518	0.516	365,034
	Apr	319,818				0.860	0.857	373,334
	May	534,441				1.187	1.182	451,996
	Jun	909,798	485,347			1.700	1.694	537,040
	Jul	1,042,919	488,288	486,818	2.142	1.935	1.929	540,767
	Aug	874,878	489,962	489,125	1.789	1.680	1.674	522,581
	Sep	685,648	492,586	491,274	1.396	1.329	1.325	517,603
	Oct	416,671	496,440	494,513	0.843	0.814	0.811	513,508
	Nov	267,402	494,145	495,293	0.540	0.582	0.580	461,326
	Dec	223,461	480,658	487,402	0.458	0.499	0.497	449,455
1981	Jan	216,905	464,955	472,807	0.459		0.475	456,978
	Feb	199,175	443,729	454,342	0.438		0.460	432,650
	Mar	219,918	416,647	430,188	0.511		0.516	426,038
	Apr	366,069	403,233	409,940	0.893	Sum of	0.857	427,323
	May	506,897	401,867	402,550	1.259	Ratios	1.182	428,701
	Jun	747,958	400,648	401,258	1.864	12.043	1.694	441,508
	Jul	854,482	400,077	400,363	2.134		1.929	443,061
	Aug	620,157	402,258	401,168	1.546		1.674	370,431
	Sep	360,670	406,721	404,489	0.892	Correction	1.325	272,274
	Oct	255,707	412,757	409,739	0.624	Factor	0.811	315,135
	Nov	251,009	415,308	414,033	0.606	0.996	0.580	433,043
	Dec	208,830	410,802	413,055	0.506		0.497	420,026

Step Two. Take a 2-month moving average of the 12-month moving average, and place the first result in the row for July 1980. This "centers" the moving average, and leaves the rows for January through June 1980 and July through December 1990 blank. The resultant moving-average value for July 1980, therefore, consists of $1/12$ the value for each month from February through December 1980 plus $1/24$ the value of January 1980 plus $1/24$ the value for January 1981, as shown in column 3 of Table 6-2.

Step Three. For each month with a moving-average value, compute the ratio of the original data series to its corresponding moving-average value. Add up all of the individual ratios for each month, and divide by the total number of months. This gives a preliminary monthly index for each month, labeled "Average Seasonal Ratios" in Table 6-2.

Step Four. Normalize the 12 monthly ratios. Determine the required normalization factor by dividing 12 by the sum of these 12 monthly ratios, which is 12.04275 in the example. Then multiply each individual monthly ratio by this normalization factor, which is 0.99645 in the example. The sum of the monthly ratios will now be exactly 12.

Each ratio shows the proportion by which water use in that month is above or below the average.

Deseasonalizing Water Use

Water use can be readily deseasonalized once the monthly ratios are computed. The deseasonalized values in column 7 in Table 6-2 were obtained by dividing each original water use value by its corresponding seasonal index. Figure 6-3 shows the 13-month moving average, the raw data, and the deseasonalized series. The moving average is much smoother than the deseasonalized data since much of the random variation in the data is removed along with seasonality, leaving only the trend component, while the deseasonalized data still include both random and trend components. Deseasonalized data can be used to examine long-term trends in water use and the variability of water use from its "normal" values.

These averaging models work well with these data, since it is from a largely built-up system that, over the period of record, did not experience rate changes in excess of the rate of overall price inflation.

The results could be less stable for a rapidly growing community, or a water system that adds new service areas. Rate developments, such as seasonal prices or overall water rate increases, also can affect the monthly water use pattern these averaging procedures seek to identify. An adaptive procedure, such as exponential smoothing, could provide superior forecasts in these circumstances.

Exponential Smoothing

Exponential smoothing,[1] as well as other extrapolation techniques, can be used to identify the trend and seasonal pattern of water use and for forecasts a few months to a few years ahead. These methods can provide reasonably accurate forecasts, as long as fundamental factors driving a time series remain stable. Accuracy can be improved by disaggregating water use by customer class.

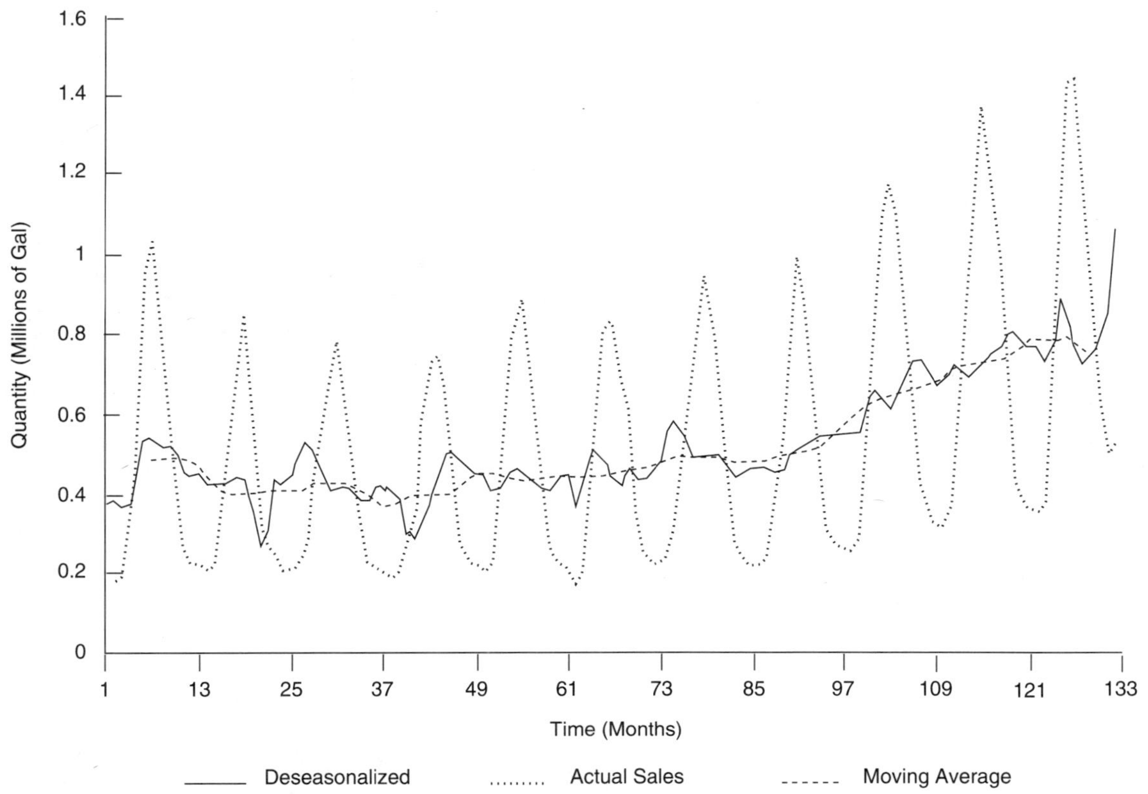

Figure 6-3 13-month, moving-average, seasonal model

A key idea is that the current level, trend, and seasonal components of a time series can be estimated as weighted averages of their past observations. Greater importance is given to recent observations by means of weights that exponentially decline for older observations.

Exponential smoothing is especially attractive for short-term forecasts when the amount of data available is quite limited and relatively few resources are available for the forecasting project.

The essential nature of exponential smoothing models will be illustrated with the simplest possible model, followed by the Winter's model, which accounts for both trend and seasonality. The simple model is concerned with removing the random variation from a series in order to produce a forecast of a constant level.

Simple Exponential Smoothing Example

Suppose we have records on the daily water use in a new commercial building. These water use records might reflect gradual leasing to full occupancy, suggesting that a simple average of the entire series would not provide the best estimate of the anticipated level of daily water use.

Denoting the forecast for the level of a series x_i $i = 1,2,\ldots,$ at time t by L_t, gives[2]

$$L_t = wq_{t-1} + (1 - w)\, L_{t-1} = L_{t-1} + w\,(q_{t-1} - L_{t-1}) = L_{t-1} + w\varepsilon \tag{6-1}$$

Where:

L = the smoothed value of the series, which is the forecast

w = the smoothing constant, $1 < w < 0$

q = observed water use

ε = the error for the forecast for period $t-1$

The first formula says that the best estimate of the future level of the series equals a simple weighted average of the current observation and last period's prediction for the level. Rearranging the terms of Eq 6-1, the final formula states that the forecast is the previous forecast adjusted by a fraction of the previous forecast error.

Choosing a Smoothing Constant

The larger the smoothing constant, the faster the forecast adjusts for the error in the preceding forecast. For a time series with a lot of random variation, the smoothing constant chosen would be relatively small. Thus, the most recent error has little impact on the new forecast. Conversely, if the series is changing rapidly, a large smoothing constant (closer to 1) would be used so that the forecast can adjust rapidly.

The smoothing constant can be set based on the analyst's evaluation of the degree to which fluctuations in the data series are composed of random variation that should be ignored. For series with much random variation, a smoothing constant of 0.1 or 0.2 is typical.[3] The constant can also be set according to some criteria, such as minimization of the squared errors over the historic series. This could be determined by trial and error using a spreadsheet or automatically by an econometric forecasting computer program.

In order to start an exponential smoothing model on a spreadsheet, the analyst must choose an initial value for L_0. This could be the most recent value of the series or an average of several historic values.

Winter's Exponential Model

Although many different smoothing models have been developed, one of the most useful is Winter's exponential smoothing model. This model is suitable for data exhibiting both trend and seasonal components.

The following discussion considers the estimation of an additive Winter's exponential smoothing model for the data charted in the solid line in Figure 6-4.[4] The dashed line indicates the exponential smoothing fit and 24-month forecast.

The data are derived from the series charted in Figure 6-1, except that the trend has been modified by creating a larger decrease and then later increase in water use in the middle of the series. The purpose of complicating the trend is to demonstrate the adaptive features of the exponential smoothing model. Exponential smoothing models allow for continual updating of the level, trend, and seasonal factors.

Updating Equations

The updating equations for the Winter's exponential smoothing model are

$$L_t = \mathrm{w}\,(q_t - S_{t-s}) + (1 - \mathrm{w})\,(L_{t-1} + T_{t-1}) \tag{6-2}$$

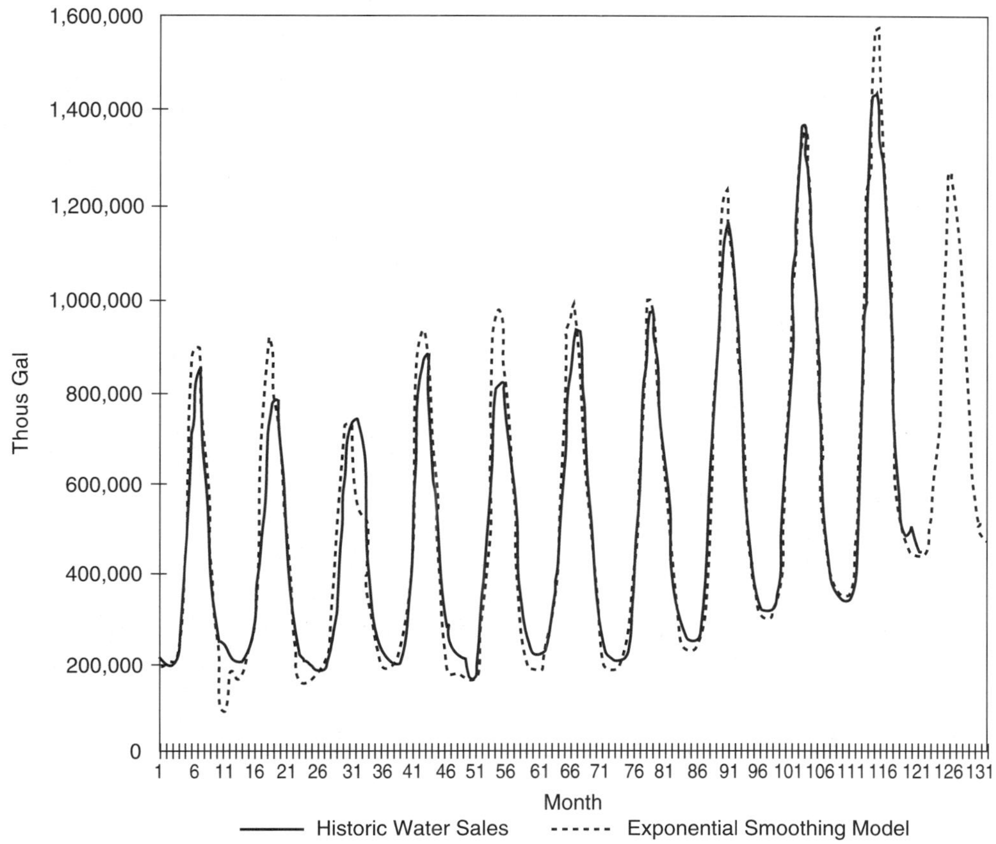

Figure 6-4 Exponential smoothing model

$$T_t = \theta\,(L_t - L_{t-1}) + (1 - \theta)\,T_{t-1} \tag{6-3}$$

$$S_t = \phi\,(q_t - L_t) + (1 - \phi)\,S_{t-s} \tag{6-4}$$

Where:

L_t = the level for period t

w, θ and ϕ = smoothing parameters for the level, slope, and seasonal components, respectively

q_t = monthly water use

S_t = the additive seasonal factor

s = 12, the seasonal length subscript

T_t = slope or trend factor for period t

The seasonal length subscript is the number of seasons per year, which in this case is 12 months. The method of calculating these components is presented in detail on the data diskette.

Choosing Smoothing Constants

The model in Figure 6-4 uses the smoothing constants w = 0.9, $\theta = 0$, and $\phi = 0.5$. These values are determined to minimize the sum of squared errors of the model for the historic monthly water use observations 1980–1990. A trial-and-error search was initiated with randomized values of the smoothing constants and computation of the associated sum of squared errors. A grid search could also have been used. The analyst also could choose the constants based on a judgmental evaluation. A large smoothing constant means that the series will adjust very rapidly to changes in the data, which is desirable if the actual data are changing rapidly. However, this also means that the model will react strongly to random fluctuations in the data. For this reason, some users of this model employ relatively small smoothing constants.

Starting Values

The initial level is taken to be the average water use in 1980. This value is substituted in for L_{12}, or December 1980. The starting value for the trend component in December 1980 is $T_{12} = 0$. A "back-of-the-envelope" trend estimate could have been used instead. The initial 12 seasonal components for 1980 are derived by subtracting L_{12} from the actual water use observations q_t. The choice of initial values is not very critical, since their effect tends to wash out within 10 to 15 time periods. Thus, the first few estimated values for the series are omitted in Figure 6-4.

For month 13, we have

$$L_{13} = 0.9(q_{13} - S_1) + 0.1(L_{12} + T_{12}) \tag{6-5}$$

$$T_{13} = 0 \tag{6-6}$$

$$S_{13} = 0.5(q_{13} - L_{13}) + 0.5 \tag{6-7}$$

Variables are as defined previously.

The calculations continue in this manner until the last period for which there is data, after which time the exponential smoothing model moves into a forecasting mode.

Forecasts

The forecasts are generated by the equation

$$\hat{q}_{t+h} = L_t + hT_t + S_{t+h-ks} \tag{6-8}$$

Where:

k = one more than the rounded down number of times h is divisible by s

Thus, the forecast for h periods hence equals the last level plus h times the trend component plus the appropriate seasonal or monthly component.

The advantages of exponential smoothing include simplicity, small data requirements, and the ability to adapt with changes in the data series. However, these adjustments to changing trend and seasonality occur with a lag, the length of which depends on the size of the smoothing constant. This prevents the model from overreacting to random variations in the data series, but also implies that it takes several iterations to catch up to a change in the environment.

Regression Approaches

The primary regression approach to identifying seasonal variation involves the use of "dummy variables." Dummy variables assume the value 0 or 1, depending on whether or not the case or observation is for a particular month or quarter. The effect is essentially similar to the additive seasonal factors in the two preceding models.

With the monthly residential water use data used in the previous averaging example, 11 dummy variables are necessary. The 12th month of December has its specific effect captured in the constant term of the regression.

The influence of temperature and precipitation is specified in terms of deviations from their average or total values for each month. Thus, average monthly temperatures for July are averaged over a long period—ideally 30 to 40 years. Then, differences between this long-term average July temperature and year-by-year average July temperatures recorded in this sample are computed. The resulting differences are used as the temperature variable. Thus, the seasonal pattern is identified with the monthly dummy variables along with the effects of higher or lower than average temperature and rainfall.

The regression can be written as

$$\begin{aligned} MWATER = \beta_0 + \beta_1\, DELTEMP + \beta_2\, DELRAIN + \beta_3\, TREND + \beta_4\, JAN + \\ \beta_5\, FEB + \beta_6\, MAR + \beta_7\, APR + \beta_8\, MAY + \\ \beta_9\, JUN + \beta_{10}\, JUL + \beta_{11}\, AUG + \beta_{12}\, SEPT + \beta_{13}\, OCT + \beta_{14}\, NOV \end{aligned} \tag{6-9}$$

Where:

MWATER = monthly residential systemwide water sales, in thous-gal units

B_0 = estimated constant

B_i, i = 1, 14 is the estimated coefficient on each variable

DELTEMP = deviation of average monthly temperature, in °F from long-term average monthly temperature for the month in question

DELRAIN = deviation of total precipitation, in inches, from long-term total precipitation for the month in question

TREND = period index for month, ranging from 1 to 132

JAN = dummy variable for January

FEB = dummy variable for February

MAR = dummy variable for March

APR = dummy variable for April

MAY = dummy variable for May

JUN = dummy variable for June

JUL = dummy variable for July

AUG = dummy variable for August

SEPT = dummy variable for September

OCT = dummy variable for October

NOV = dummy variable for November

Dummy variables assume a value of 1 when the water use observation pertains to their month. Otherwise, the dummy variables are assigned a value of 0. A dummy variable for one of the 12 months—in this case, December—is not needed and would conflict computationally with specification of the constant term indicated by β_0 in Eq 6-9. β_0 is named CONSTANT in the printout from the spreadsheet regression shown in Table 6-3. Each monthly dummy, therefore, shows the difference between that month's water use and the "base" level of water use in the equation—that of the omitted month of December. The choice of which month to omit is entirely arbitrary and has no effect on the results.

A simple, linear time trend is allowed for in the regression equation by including the explanatory variable *TREND*. The value of this explanatory variable starts at 1 for the first month in the historic record in 1980. The variable increases by 1 for every successive month in the data. Alternatives to this simple linear trend, such as population, could be specified.

The data diskette presents the complete data on *MWATER*, *DELTEMP*, *DELRAIN*, *TREND*, and the dummy variables, so the reader can follow the discussion, derive the results, and explore alternatives.

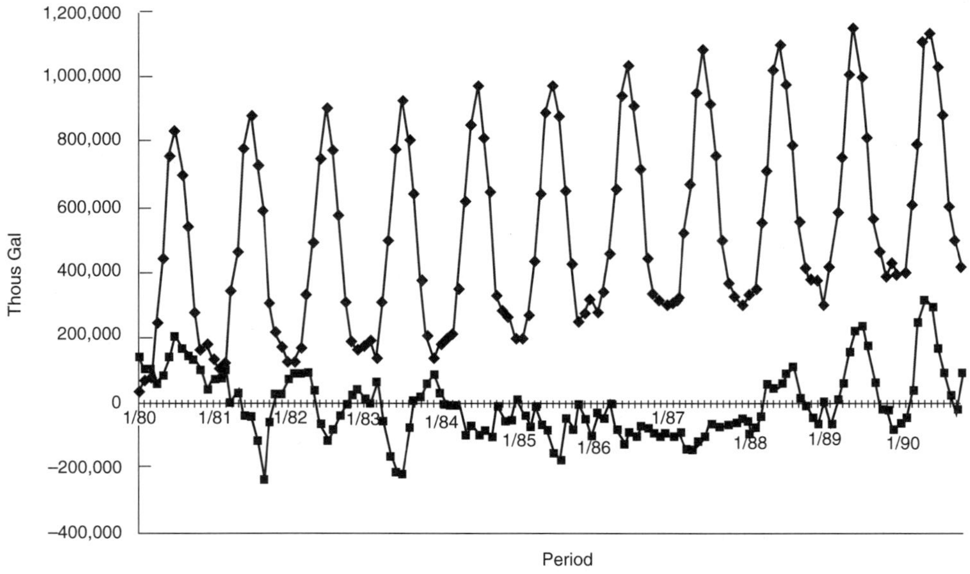

Figure 6-5 Seasonal regression and plot of residuals

Graph of the Seasonal Regression

The results of the regression are graphed in Figure 6-5 together with the residuals. The regression equation is the top curve in the figure. The regression curve tracks the seasonal rise and fall of total residential water use and the long-term trend in this system relatively well.

Since the estimated trend is linear, the regression misses the changes in the trend. This results in numerous positive residuals near the beginning and end of the series, and negative residuals during the middle years. The use of nonlinear trends, such as those described in chapter 4, would have avoided these runs of positive and negative residuals. However, the use of a nonlinear trend would only improve the forecast derived from this model if that same nonlinear relationship were to continue in the future.

The residuals are shown in the lower curve in the figure, hugging the horizontal axis. The residuals are calculated by taking the actual water sales for a given month and subtracting the water sales predicted by the regression equation.

Coefficients of the Seasonal Regression

The coefficients estimated by this regression are listed in the lower panel of Table 6-3. Their signs match our intuition in many cases. Thus, the coefficient of *RAIN* is negative, indicating that system water demand is reduced with higher rainfall. Higher average monthly temperatures, on the other hand, tend to increase monthly water demand. Consistent with the gradual upward slope of the smoothed demand series, the coefficient of *TIME* is positive and statistically significant. The coefficients of the dummy variables *JAN* through *NOV* identify the relative pattern of monthly water use, which suggests that peak water use occurs in July.

Predicted water use for each month is equal to the constant of the regression (72,952) plus the coefficient from the specified month dummy variable plus the values of *DELTEMP*, *DELRAIN*, and *TREND* times each of their estimated coefficients. The predicted water use for July of year 2, when *DELTEMP* = 1.81, *DELRAIN* = –1.12 in., *TREND* = 19, and *JUL* = 1, is

$$889{,}533 \text{ gal} = 72{,}952 + 5{,}910 \times 1.81 - 10{,}579 \times (-1.12) + 2{,}859 \times 19 + 739{,}715 \times 1$$

Similar calculations would give predicted water use for each month and year.

An important end result of the estimates presented in Table 6-3 is the opportunity to identify peak to average month relationships.[5] Substituting in 0 for *DELTEMP* and *DELRAIN* and the midpoint of the values driving *TREND* (66), the model suggests a peak-month to average-month demand ratio of approximately 1.9.

With disaggregated data on individual household water use by month, the modeling problem is more complex, but richer in implications. Thus, seasonal residential water demand models often are developed for summer and winter, since the requirements for outdoors water use can vary so much between these times of the year. Such regression models inclusive of a price term or terms provide insight into the effect of policy measures, such as peak demand pricing.

Many researchers would put population or number of water customers into the demand model in place of the simple additive time trend to reflect varying growth rates. A complete model would include additional variables, such as price and conservation measures.

The short-run forecast can be adjusted toward "real time" by including current weather forecasts in the model. Current month and year-to-date water sales could be included as leading indicators of water demand, further improving forecast accuracy.

Table 6-3 Summary output: seasonal residential water use equation

Regression Statistics	
Multiple *R*	0.94
R^2	0.89
Adjusted R^2	0.88
Standard Error	108,382.12
Observations	132

Analysis of Variance

	df	SS	*MS*	*F*	Significance
Regression	14	1.13769*E*+13	8.12639*E*+11	69.18	1.0189*E*–49
Residual	117	1.37436*E*+12	11,746,683,368		
Total	131	1.27513*E*+13			

	Coefficients	Standard Error	*t* Stat	*P* Value	Lower 95%	Upper 95%
CONSTANT	72,952	37,278.02	1.96	0.05	-875.19	146,779.12
DELTEMP	5,910	2,782.38	2.12	0.04	399.73	11,420.45
DELRAIN	-10,579	11,584.46	-0.91	0.36	-33,521.71	12,363.10
TREND	2,859	249.14	11.48	0.00	2,365.95	3,352.76
JAN	-9,497	46,295.48	-0.21	0.84	-101,182.92	82,188.59
FEB	-17,968	46,281.40	-0.39	0.70	-109,626.03	73,689.72
MAR	6,976	46,268.65	0.15	0.88	-84,656.40	98,608.88
APR	181,274	46,257.25	3.92	0.00	89,663.49	272,883.60
MAY	360,452	46,247.18	7.79	0.00	268,862.12	452,042.35
JUN	636,623	46,238.46	13.77	0.00	545,050.18	728,195.86
JUL	739,715	46,231.08	16.00	0.00	648,156.29	831,272.72
AUG	605,796	46,225.03	13.11	0.00	514,249.78	697,342.28
SEPT	428,156	46,220.33	9.26	0.00	336,618.94	519,692.83
OCT	161,045	46,216.98	3.48	0.00	69,514.55	252,575.14
NOV	38,555	46,214.96	0.83	0.41	-52,971.70	130,080.91

Summary

The seasonal pattern of urban water use is determined by a variety of factors, the most important of which is likely to be irrigation demand. A causal regression model with temperature, precipitation, and monthly dummy variables, such as presented in this chapter, illustrates one method that can capture the seasonal pattern of demand.

Simpler averaging and univariate approaches show promise in predicting seasonal water use ratios. The averaging procedure described at the beginning of the chapter produces reasonable estimates of the time pattern of water use when other important influences, such as water commodity charges, are relatively constant.

The Winter's exponential smoothing model can achieve a close fit to an actual water use series containing systematic seasonal variation and trend. Exponential smoothing models are adaptive in the sense that they supply updated estimates of level, trend, and seasonal factors each time additional historic observations are supplied. These models may be suitable for short-term forecasting and analysis of seasonality. Exponential smoothing also may have useful applications for the prediction of related variables, such as growth in the numbers of various housing types, where data series can be short in length and reflect sharp changes in activity over periods of six months to a year.

In many contexts, short-term forecasting benefits from a type of momentum that seems to be associated with the direction of things taking place in the present. In intuitive terms, momentum is the idea that whatever has been happening is likely to continue a little longer. However, when key factors influencing water use change, such as prices and incomes, peak demand relationships also can change, and multivariate models provide more accurate forecasts.

Notes

1 Exponential smoothing models evolved in an applied business forecasting context. Subsequent research has shown how they can be viewed as solutions to certain types of formal time-series models known as autoregressive integrated moving-average (ARIMA) models. Box–Jenkins or ARIMA models are based on the theory of stationary stochastic processes, and are estimated with curve-fitting techniques following three steps

1. Model identification.
2. Model estimation.
3. Diagnostic testing.

In the Box–Jenkins methodology, time series are viewed as being built up by autoregressive and moving-average components. It is often necessary to compute first or higher differences of a time series, to render it stationary (constant) in its mean, and to take various transformations of the series to render it stationary in its variability over time. Once a time series is transformed to a stationary series, analytic tools, such as the autocorrelation and partial autocorrelation functions of a time series, can be used to identify the types and mixture of autoregressive and moving-average processes present.

After the model is identified, its parameters usually are estimated by methods of numerical approximation. Fully specified ARIMA models usually require more historic data to estimate than exponential smoothing models.

[2] If this initial equation is projected backward in time, the forecasts for period t and $t-1$ can be written as

$$L_t = w\, q_{t-1} + w(1-w)\, q_{t-2} + w\,(1-w)^2\, q_{t-3} + \ldots$$
$$L_{t-1} = w\, q_{t-2} + w(1-w)\, q_{t-3} + w\,(1-w)^2\, q_{t-4} + \ldots$$

Multiplying both sides of the second equation by $(1-w)$ and subtracting gives the recurrence form of the simple exponential smoothing procedure presented as Eq 6-1.

[3] Thus, if one believes that the past plays a relatively minor role in patterning a series, the smoothing constant can be set close to 1. Values of w close to 0, on the other hand, accord greater importance to the range of past observations on a variable.

[4] The Winter's model can also be estimated with the trend and/or seasonality in multiplicative form.

[5] Serial correlation or autocorrelation exists in the residuals of the equation in Table 6-3. The primary impact of this autocorrelation is on the standard errors of the estimated coefficients. The point estimates indicated by the coefficients in Table 6-3 should be unbiased if specification error is not a factor. Therefore, it is acceptable to use the unadjusted regression estimates to estimate peak to average relationships. (See Pindyck and Rubinfeld 1991.)

References

Pindyck, R.S., and D.L. Rubinfeld. 1991. *Econometric Models and Economic Forecasts*. 3rd ed. New York: McGraw-Hill.

Chapter 7

Population, Economic, and Technological Forecasts

Water planners and forecasters must evaluate and use forecasts of population, economic growth, and technology produced by others. This chapter reviews some basic facts and guidelines that may be useful for such evaluations.

Population, economic growth, and technical change are "big-picture" factors that, over long periods, account for profound changes in patterns of urban water use. Over the last century, the United States has gone from a predominantly rural to an overwhelmingly urban nation. Population quadrupled from around 65 million to more than 250 million persons. With mass production of the automobile in the 1920s, urban land use patterns began a transformation that accelerated after World War II and continues to this day. The central city is now surrounded by burgeoning suburbs, whose residents often live a "water-intensive" life-style, complete with landscaped yards and a full complement of water-using appliances.

Just as we have become accustomed to this pattern of growth, there are signs of change in the average age of the population, job opportunities, and "technical fixes" that have the potential to save water.

National Population Forecasting

The population for the United States is generally forecast in a highly disaggregated manner. Not only is population change broken down into births, deaths, and migration, but also by age group (cohorts), sex, and ethnic/racial groupings.

The current US Bureau of Census population forecast for the United States estimates projected fertility rates, projected survival rates, future net immigration, and armed forces overseas population for 16 different race/ethnic/sex matrixes with a cell for each year of age 0 to 100 and over (Day 1993). The ethnic/racial groupings include White, Black, American Indian, Eskimo and Aleut, Asian and Pacific Islander, and Hispanic origin. Fertility and mortality rates differ markedly between some of these groups.

National forecasts are derived by combining numerous individual forecasts for ethnic/ racial groups, taking into account the age and sex structure.

Since longevity by age group generally changes very slowly, it is relatively easy to predict the size of the 60–70 age group in 10 years, given the current size of the 50–60 age cohort. Thus, we might expect population forecasts to be relatively accurate. Predicting national birth rates has proven difficult, however, as documented by Ascher (1978) and Stoto (1983). Considering forecasts of the US population from 1971–1981, Ascher tabulates 20–30 percent errors in predicted national population growth rates 5 to 30 years after a projection is put forward.

Because of these recurrent errors in their forecasts, the Census Bureau offers long-range population forecasts with high, medium, and low growth scenarios as shown in Figure 7-1 for a forecast published during 1993.

Population growth is closely linked with job availability, but has a dynamic of its own related to adjustments in fertility, mortality, and the age structure.

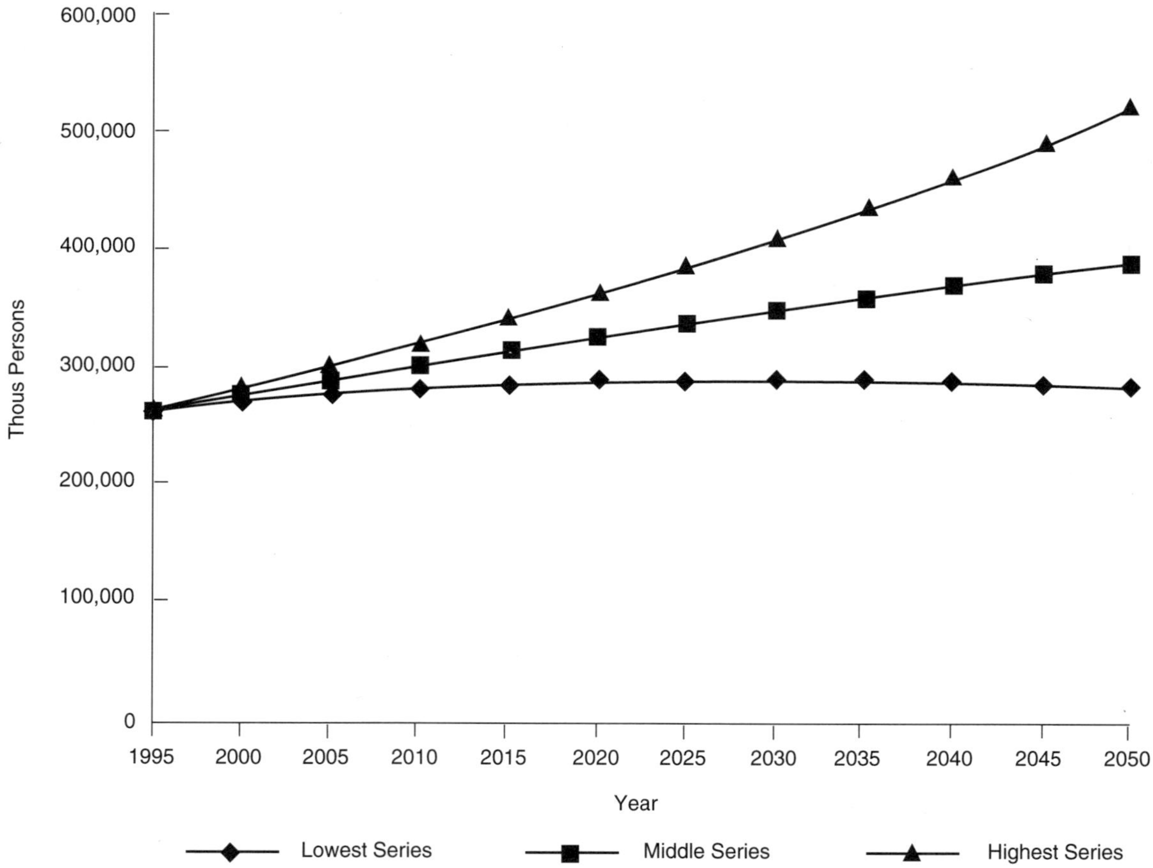

Figure 7-1 US population projection, US Census Bureau

In the decades ahead, population growth in the middle and lower forecast series in Figure 7-1 is predicted to slow, as the American population ages. On the other hand, immigration and higher birth rates among select groups in the population may take up some of the slack.

Population Trends

Major shifts in recent demographic patterns include

- gradually slowing population growth
- increases in the median age of the population
- continuing reductions in household size
- growth in single-parent households and households composed of unrelated persons

Population growth during the decade of the 1980s was 9.8 percent. This was the second lowest percentage growth for a decade in US census history. Only the decade of the Great Depression exhibited lower overall population growth. By contrast, US population growth reached 18.5 percent over the decade of the 1950s, which included the peak of the "baby boom" (1946–1964) and an average of more than three births per woman.

Over the 1990s, continuing declines in the rate of US population growth are forecast. The annual growth in population is anticipated to decrease from about 0.8 percent to 0.6 percent by the end of the decade. The primary reason for this trend is a decline in fertility. Many factors appear to contribute to this decline in childbearing, including increased numbers of women in the workforce, postponed childbearing, the availability of effective contraception, changing marital patterns, and declining job prospects and earnings of young people. Projections of this trend, produced in the early 1990s by the Census Bureau, suggest that the US population will stop growing sometime between 2020 and 2030.

Economic integration of North, Central, and South American markets may forestall this plateauing of US population. Recent long-range forecasts from the Census Bureau emphasize the possibilities of higher birth rates among new immigrant and specific ethnic groups. As the Hispanic population increases, these higher birth rates may exert growing impacts on the US population. Also, with increased trade may come accelerated movement of people.

Regional Population Trends

Two major trends impacting regional populations in the United States are continuing reduction in the population of core city areas and expansion of peripheral suburban areas or cluster dwellings for retirement. Locations in the South and West continue to grow rapidly. The five states showing the highest percentage increases in the 1990 census, for example, were Nevada (50.1), Alaska (36.9), Arizona (34.8), Florida (32.7), and California (25.7). The 1990 census revealed that 9 out of the 11 most rapidly growing metropolitan areas over the decade of the 1980s were in Florida. Naples, for example experienced a 77 percent increase in population over the decade. The economic distress experienced by California in the early 1990s, however, demonstrates how rapidly these trends can change.

Other areas have experienced population decline. Older core urban areas, such as Boston, New York, and Denver, lost population over at least two censuses. The reasons for population loss are complex and can involve flight to the suburbs, loss of employment base, and regional economic decline.

These shifts suggest that simply extrapolating the trends of the past few decades can lead to large errors in forecasting.

Regional Population Forecasts

Regional and local population forecasts are far more important than national forecasts for local utility planners. While errors in birth rates plague national population forecasting, errors in migration assumptions frequently generate even larger errors for local and regional forecasts.

Regional population forecasts can be based on the following approaches:

- **extrapolation** of the current average growth rate, perhaps including some adjustment factor throughout the forecast period
- **structural models** using causal variables in which population growth is the lefthand side (dependent) variable whose variations are "explained" by a set of independent variables representing social and economic conditions
- **land-use ratios**, which allow analysts to infer the eventual buildout population from the land use plan and zoning ordinances
- **economic-base models**, which link population growth to growth in economic base industries and employment
- **cohort-component techniques** based on assigning different fertility, migration, and death rates to various age groups or cohorts (Pittenger 1976)

All such approaches work with imprecise base population numbers, except just after a census. Methods for estimating the existing population in a noncensus year draw on information from county and state birth and death records, as well as information about telephone, electric, and water hookups, school enrollments, and so on. Even the decennial census contains errors and omissions, which are likely to be larger in central city areas where population densities are high and employment rates can be low.

The most important factor affecting population change in a region is usually the economic situation. Employment opportunities are a magnet for new migration while also supporting family formation and the natural increase of the population. Depopulation is often experienced by individual communities as well as entire states (e.g., Wyoming, Idaho, and the Dakotas) lacking job growth.

A Checklist

Given both the need for and unreliability of long-range population projections, utility analysts should adopt a "due diligence" approach to using population forecasts. A population forecast meets the due diligence test when the answers to the following questions are "yes."

- Does the population forecast offer high, medium, and low variants?
- Does the population forecast justify the estimate of the existing population with reference to information gathered from sources including birth and death records, school enrollment, and utility connections?
- Is the forecast consistent with other regional and national forecasts?
- Does the population forecast acknowledge developments such as aging of the population and associated drops in the fertility rate?
- If not, does the population forecast explain why these commonly perceived trends will not be operative in the utility service area?
- If the population forecast is for a smaller population area, does the forecast acknowledge the relatively higher potential for variability?

When population forecasts do not meet these criteria, they may be more subjective estimates on the part of the agencies, organizations, or persons producing them. Recognizing this, utility planners should develop their own population growth scenarios. These scenarios can be the product of brainstorming sessions and are useful in exploring the sensitivity of capacity expansion plans.

Other Demographic Factors

Total population is not necessarily the whole story for purposes of understanding water use. Other demographic features, such as household size and, possibly, the age structure, can significantly influence residential water use.

Studies have established, for instance, that there are economies of scale in household water use. Two one-person households use as much as 60 percent more water than one two-person household. Thus, the trend toward smaller families and more single individuals can contribute to increases in water use per capita, providing housing patterns remain relatively constant.

Economic Forecasts

Economists have been ridiculed about their disputes and pretensions. George Bernard Shaw once said, “If all the economists in the world were laid end to end, they still could not reach a conclusion.”

Yet longer-range private and public sector economic forecasts, ranging from a few to 20 years, often look quite similar. These forecasts usually project an inflation series, such as the consumer price index (CPI), and production and income quantities, such as gross domestic product (GDP), personal income, and disposable income.

There are private services that allow on-line access to forecasts and associated databases on an annual subscription basis. There are also long-range economic forecasts maintained by agencies of the federal government. The Bureau of Economic Analysis (BEA) in the US Department of Commerce, for example, produces long-range economic forecasts for the nation, each state, and each county (US Department of Commerce

1990). The forecast, which is scheduled for revision every five years, includes employment and earnings by industry and population. The BEA forecast uses a "step-down" method, in which national economic growth is forecast based on labor projections from the Bureau of Labor Statistics (BLS) and other data. The BEA then allocates these national projections to individual states and industrial groupings.

Accuracy Issues

Although various long-range economic forecasts often tell basically the same story, these forecasts are subject to errors over time comparable to population forecasts. This is not surprising in view of linkages between the economic and demographic factors.

Typically, the smaller the region or area of the forecast, the larger the expected error. This is because some errors net out in predictions of total population or economic activity.

Since 1968 the American Statistical Association and the National Bureau of Economic Research have tracked about 50 economic forecasting operations. This analysis demonstrated that predicting turning points, i.e., the onset of recessions and business upturns, is extremely difficult. Additionally, the errors in the forecast of the annual growth rate of GDP are typically as much as 33 percent. That is, if the growth rate were 3 percent of GDP, the typical forecast error would be 1 percent of GDP.

The uneven performance of economic forecasting models was especially obvious after the energy crisis of the early 1970s. This prompted much reevaluation by analysts. One upshot was a now-famous forecasting competition organized by Spyros Makridakis. The "M-competition," as it is known in forecasting circles, produced interesting findings, not all of which flattered the forecasting profession. One conclusion was that simple forecasting techniques often work about as well as complex ones, especially for longer-range forecasts. The evidence suggested a slight advantage may exist with regard to exponential smoothing (discussed in chapter 6). All the techniques studied lost explanatory power or accuracy after about five periods out in their forecasts regardless of whether the model used weekly, monthly, quarterly, or annual data (see Makridakis 1986 and Ashley 1988).

A Checklist for Regional Economic Forecasts

The assumptions underlying economic forecasts should be scrutinized, since faulty assumptions will result in unreliable forecasts. The following questions or issues might be considered when analyzing economic forecasts:

- What industries and sectors fueled growth in the past?
- Are these industries likely to continue playing an important role in the future, or do they belong to "declining sectors"?
- If growth has been led by firms producing goods for sale outside the area, is the national or international market likely to continue growing?
- Are local producers competitive?
- If growth has been supported by in-migration, are the factors that made the region desirable still in place?
- If the region has lost major employers or industries, is there sufficient size and complexity in the local economy to attract new firms and industries?

Answering these questions pushes the analysis in the direction of scenario construction. Before striving for quantitative precision, the planning community can sit down and attempt to identify plausible scenarios about the future. Internal consistency and linkage with currently perceived and emergent tendencies are the primary criteria for good scenarios. Scenarios can be valuable in clarifying how long-term water projects may play out under the variety of conceivable future conditions.

Personal Income

Many studies demonstrate a positive relationship between water use and personal income per capita or per household. From analysis of 27 different statistical studies, it appears that the average increase in urban water demand resulting from a 10 percent increase in real income is 4.5 percent. Although the analyst would generally prefer to use income per household rather than income per capita in a forecasting model, this variable is not always readily available.

Although this response is less than one-for-one, the cumulative effect can be significant. For example, if real income per capita (after adjusting for inflation) rises by 2 to 3 percent per year (as it did after World War II), per capita water demand can be expected to increase by 20 to 30 percent over 20 years. Ignoring the impact of such changes clearly introduces significant errors in the long-term water demand forecast.

Water use goes up with increases in real income because additional water-using items, such as clothes washers, dishwashers, garbage disposals, luxurious landscaping, and swimming pools, are affordable. In addition, as incomes rise, people may be less concerned with their water bills, since these bills typically absorb a smaller proportion of their income. If these "high-income" water uses approach saturation in a community, further income increases may have much less effect on water use. When the use rates of these water-using home features are included in the specification of a demand model, the size of the income elasticity of demand will be dramatically diminished. The acceptance rates of these water-using features frequently are highly multicollinear with income, and are therefore omitted from a demand model specification. In this case, the income variable represents the use rates of these home features.

Effect of Inflation

It is appropriate here to stress the distinction between *real* versus *nominal* variables in economic parlance. In the United States, the inflation rate is computed by the BLS against a market basket of goods and services that consumers typically buy. Prices for this group of goods and services are sampled across the nation periodically. The BLS then computes the CPI. Increases in the CPI indicate inflation or increases in the level of prices. In other words, it takes more money to buy the same market basket of goods when the CPI increases.

A consumer may take account of how much the purchasing power of his or her income is reduced by inflation. For such consumers, their real, or inflation-adjusted, income is the key factor in their buying decisions, rather than their nominal income, i.e., the number of dollars they are paid as a wage, salary, or other compensation for services rendered.

Income Trends

It is essential to examine individual community characteristics and the pattern of change occurring in the utility service area to evaluate the potential for income growth. Both the number of jobs and average salary levels are important. The replacement of high-paying manufacturing jobs with low-paying service jobs, or its reverse, would obviously change community water demands if this change occurred on a sufficiently large scale.

Whether real per capita income will continue to increase at its historic rate of between 2 and 3 percent is an open question. Real per capita incomes have increased over the past two decades largely because of the emergence of the two-wage-earner household. However, there have been declines in the purchasing power of the average paycheck simultaneously with the increased labor force participation rate. There are limits to this compensation process, marked by full (100 percent) labor force participation of women. Based on the growth in two-wage-earner households, real income improvement is likely to plateau. Remember that local area income trends are of primary importance to the utility analyst, and these may vary substantially from the national trend.

Technology Forecasts

One challenging problem facing the water forecaster is to project the acceptance, acquisition, or assimilation of water-saving and water-using technologies, appliances, conveniences, and devices. There are products on the drawing board or in prototype today that have the potential for substantially affecting future urban water demand. For example, there has been discussion of genetic engineering of extremely low water-demanding lawns that are comparable in appearance and touch to existing lawn grasses. In some parts of the country, increasing numbers of hot tubs, saunas, and swimming pools are influencing water consumption.

The acquisition or penetration of specific products in a community is a *diffusion process*. Just after introduction of the product, there may be a period of slow acceptance. When 10 or 20 percent of the population acquire the good in question, its acquisition rate is likely to accelerate. After a higher proportion of the population acquires the product, the acquisition rate slows again, suggesting a type of saturation phenomenon. Eventually, market penetration reaches a more or less stationary plateau (not necessarily at 100 percent). In the case of costly durable goods, such as home appliances, inadequate income or housing space provides primary barriers to acquisition for many people. Graphically, the diffusion process results in an S-curve.

The trick in predicting the acceptance rate of some new water-using or water-saving appliance or device is to figure out the shape of the S-curve over time. The data diskette shows how this S-curve, or sigmoid curve, bends and changes slope as the underlying diffusion parameters are modified. Expert judgment plays the major role in establishing a reasonable value for such diffusion parameters. Methods such as the Delphi process (see appendix A) formalize the process of recording these judgment calls and continue until a convergent estimate of acceptance rate and penetration is reached.

Technology and Industrial Water Use

One major trend in water demand has been the increasing efficiency of industrial water use. Much of the industrial water used is self-supplied, but a common problem of industrial users of potable and self-supplied systems is treatment costs associated with effluent standards.

Water use coefficients or patterns of specific industries vary widely from place to place, from industry to industry, and from firm to firm.

For individual firms granted special industrial water use rates, a condition of service could be the provision of periodic water use forecasts by the firms to the water utility to aid in system planning. These multiyear forecasts aid the utility in preparing for changes in industrial demands and help to avoid either excessive or insufficient system capacity.

Space cooling generally uses a large amount of electricity and water (both directly and indirectly through the generation of electricity) in all sectors of the economy. Such water use is primarily a function of the size of the cooled space and the climate. The technology involved generally evolves rather slowly; there is no reason to expect sudden changes. However, there is always the possibility of a cost-reducing technological breakthrough that may be either water saving (e.g., air-cooled compressors) or water using (e.g., multistage evaporation).

Appliances

The proportion of households with water-using appliances is only partly a function of income. It is also related to cultural factors, technological improvements, housing characteristics, and the price of the appliances.

An American Water Works Association survey (Solley and Pierce 1988) noted the following:

> During the period from the late 1940s to the early 1960s, a pronounced change occurred in water use habits of the US public. In the years following World War II, average residential water use per customer saw a steady climb as automatic clothes washers, dishwashers, garbage disposals, and other water appliances were introduced and accepted by the public. The construction boom in single-family homes also promoted residential water use as new households watered and cared for their new and usually large lawns. Growth of the two-car family status increased domestic as well as commercial auto-washing requirements, and widespread use of evaporative air conditioning in western areas also contributed to increased water demands.

While some appliances, such as water softeners, garbage disposals, reverse-osmosis water filters, and evaporative coolers, increase community water use, other appliances, including clothes washers and dishwashers, do not necessarily increase total water use. Following is a brief review of appliance characteristics in regard to water usage.

Clothes Washers

Clothes washers are an obvious convenience. Since the wash must be done one way or another, home washing machines may or may not change overall community water use. Their water use efficiency, however, has increased markedly in recent years.

Dishwashers

Dishwashers may either increase or decrease household water use, depending on dishwashing habits. For those who prewash dirty dishes before putting them into the dishwasher, total water usage is increased. For people who would otherwise wash everything under running water, but simply put it in the machine, the use of the machine saves water.

It should be noted that a machine with a water-conserving option does not guarantee that people will use it. Many people believe that reducing water use will increase the chance of leaving some dishes imperfectly cleaned.

Garbage Disposals

The use of garbage disposals in sinks increases household water use because it replaces a nonwater-using disposal method, i.e., putting garbage in the trash sack, with a water-using technology. Individual household choices of using or not using the disposals may be significantly impacted by community mores and regulations. Some communities may prohibit the use of disposals, while others may effectively require their use by prohibiting the disposal of wet garbage in the regular trash.

Community efforts to encourage or mandate recycling of metal cans, glass bottles, etc., may also have an impact on the use of disposals. If the frequency of trash collections is reduced as part of a program to institute the frequent pickup of recyclable materials, households that previously disposed of garbage in the trash may switch to disposals. Specifically, if a community switches from twice-weekly trash collection to once weekly, some water customers may install disposals to reduce the accumulation of garbage and its odor.

Trash-reduction programs that institute a charge per can of trash pickup may similarly encourage the use of disposals as households attempt to reduce both the volume and frequency of the service used.

Reverse-Osmosis Water Filters

As households, particularly those with higher incomes, express increasing concern about the quality of their drinking water, the use of water filters is likely to increase. The most efficient current technology to remove dissolved chemicals appears to be reverse osmosis, which produces a stream of slightly brackish wastewater. Reverse osmosis uses 5 to 10 gal of water for each gallon of drinking water produced.

Water Softening

Conventional water softening produces wastewater with a high salt content. Not only does this cause some incremental increase in water use, but the salt added to the wastewater reduces its value for reuse.

Evaporative Coolers

Many homes and small businesses in hot desert climates use evaporative coolers as a lower-cost alternative to air conditioning. Unlike the other appliances discussed, evaporative (swamp) coolers have a negative income elasticity of demand. As incomes rise, households have a tendency to substitute the much more expensive electric air conditioning. This substitution, however, is sensitive to electric rates, leading higher income households to install dual systems that use coolers when the relative humidity is low and air conditioning when it is high.

In areas where evaporative coolers are commonly used, their falling use with rising incomes provide an offset to increased use of water for other purposes as incomes rise.

Summary

The planning horizon for water supply systems often is several years, even up to one or two decades. This represents a period of time sufficient to plan and implement large-scale and complex facilities. Planning in the long term requires analysts to forecast population, economic activity, and technical conditions decades into the future.

Despite their importance, evidence suggests the accuracy of such forecasts is relatively low. Water resource planners must evaluate population and economic forecasts relevant to future water use. They must determine if the forecasts are merely straightline extrapolations or if they try to come to grips with the underlying processes at work.

A checklist to help evaluate population and economic forecasts is provided in this chapter, but the specific questions in the checklist are provided as suggestions only. The point is that the future for many communities may vary from historical trends. This puts more responsibility on planners to probe the assumptions of long-range forecasts necessary to infrastructure expansion and maintenance.

The checklist approach can be supplemented with scenario development. Instead of striving for quantitative forecasts, combine anticipated elements in a plausible "story" about the future. A reasonable scenario must be internally consistent and have plausible points of attachment with current trends. Scenarios can be valuable in clarifying how long-term water projects may play out under the variety of conceivable future conditions.

In the case of water utilities, too, scenarios for the future can and should focus quite specifically on the complex linkage of socioeconomic and urban land use factors influencing water consumption. For example, urban residential land use ranges from low-density areas with single-family houses on large lots, to medium-density areas with smaller single-family lots and some multifamily units, to high-density development consisting of larger apartment or high-rise complexes.

While indoor water use depends primarily on household size and income, outdoor use depends on climate and the local consensus regarding appropriate landscaping choices. A shift from single-family housing on large lots to apartment or townhouse living will reduce water use per capita, but may increase water use per square mile, as population densities increase.

Less dramatically, a gradual shift to smaller urban lots or the development of a social consensus that discourages lush green landscaping may reduce water use both per household and per acre. Conversely, in an area with rising incomes, the social consensus may push strongly toward larger lots with more extensive landscaping and, as a result, increasing outdoor water use.

Sorting out the exact balance of these factors for periods some years into the future is difficult, if not impossible. It is, however, extremely constructive to conceptualize how the various elements might fit together.

References

Ascher, W. 1978. *Forecasting: An Appraisal for Policy-Makers and Planners*. Baltimore, Md.: The Johns Hopkins Press.

Ashley, R. 1988. On the Relative Worth of Recent Macroeconomic Forecasts. *International Journal of Forecasting*, pp. 363–376.

Day, J.C. 1993. *Population Projections of the United States, by Age, Sex, Race, and Hispanic Origin: 1993 to 2050*. Current Population Reports P25-1104. US Department of Commerce, Economics and Statistics Administration, Bureau of the Census.

Makridakis, S. 1986. The Art and Science of Forecasting: An Assessment and Future Directions. *International Journal of Forecasting*, pp. 15–39.

Pittenger, D.B. 1976. *Projecting State and Local Populations*. Cambridge, Mass.: Ballinger.

Solley, W.B., and R.R. Pierce. 1988. Trends in Water Use in the United States, 1950 to 1985. In *Water Use Data for Water Resource Management, Proceedings of a Symposium*. Bethesda, Md.: American Water Resources Association.

Stoto, M.A. 1983. The Accuracy of Population Projections. *Journal of the American Statistical Association*, 78 (March 1983): 13–20.

US Department of Commerce. 1990. *BEA Regional Projections to 2040*, 2 volumes. Bureau of Economic Analysis, US Department of Commerce.

Chapter 8

Weather and Climate

In this chapter, *weather* refers to short-term fluctuations in temperature, cloud cover, humidity, precipitation, and wind conditions. *Climate*, on the other hand, refers to longer-term recurring variations in temperature, precipitation, and wind, including seasonal patterns. Climate is also used to refer to multiyear drought or flood cycles and secular or very long-term trends in temperature or rainfall.

Seasonal components in water use are generated largely by the local climate. Summer peaking demand is typically related to water use for outdoor activities, including lawn irrigation, gardening, and use of swamp coolers. Conversely, some communities in Alaska have winter-peaking systems because water flow is used to prevent the freezing of pipes. Seasonal demand patterns are important in establishing planning numbers, such as the peak to average demand and maximum day demand, which guide the sizing of water treatment and distribution systems, as well as wastewater systems.

Weather, on the other hand, contributes to volatility or changeability of water use during seasonal cycles. Weather is difficult to predict, especially for more than a few days into the future. Studies conducted by the National Weather Service confirm that weather forecasts are most accurate for a few hours up to a day or two. Longer-range weather forecasts focus on broad features of regional weather patterns, rather than specific numerical predictions (see Wagner 1989 and National Weather Service 1991).

Precision in water demand forecasts based on weather factors, accordingly, is limited to very short-range applications—a matter of hours or a few days. Such forecasts might be used to fine-tune operations of pumps and reservoir gates.

Otherwise, weather and climate are important in assessing the potential variability of water demand and contribute to various types of risk analysis.

This chapter focuses on the variability of temperature and precipitation while also considering other weather-related variables, such as evapotranspiration. The chapter concludes with a brief discussion of long-term climatic trends, including heat-island effects and hypothesized global warming.

Correlation of Temperature, Rainfall, and Water Use

The role of climatic factors in seasonal variations in water use can be investigated with data shown in Figures 8-1 and 8-2. These charts show more than a decade of weather information for the same city, the aggregate residential water sales of which are presented in Figure 2-1 (on page 23). Note the interrelated cyclical pattern in these three graphs.

Correlation Coefficients

The *correlation coefficient* measures the similarity of variation of two variables. Correlation coefficients range from +1 (which indicates a perfect linear relationship between the two variables), to 0 (which indicates absolutely no relationship), to −1 (for a perfect inverse linear relationship). A correlation coefficient between 0 and +1 means that the variables increase or decrease together, but not identically.[1]

It is difficult to be sure of the correlation between water sales, temperature, and precipitation by simply looking at the graphs, although water sales and temperature appear

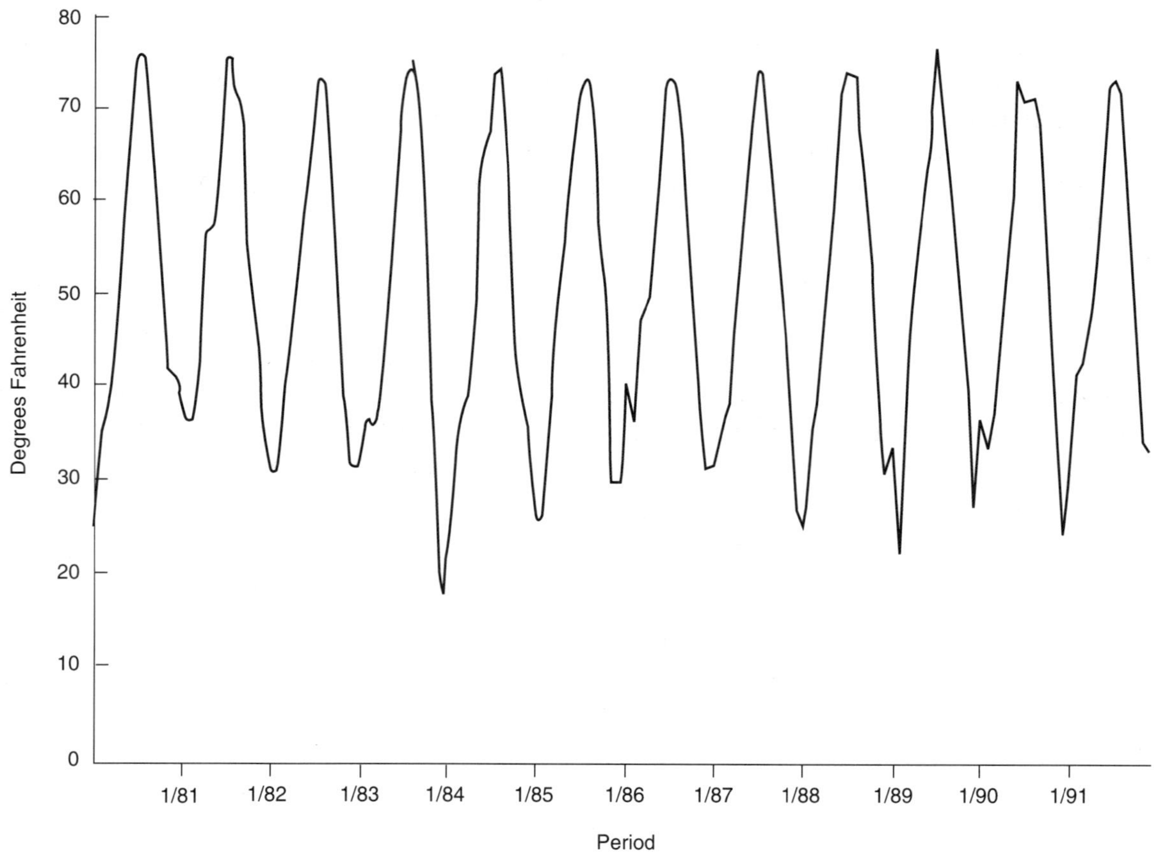

Figure 8-1 Monthly average temperature

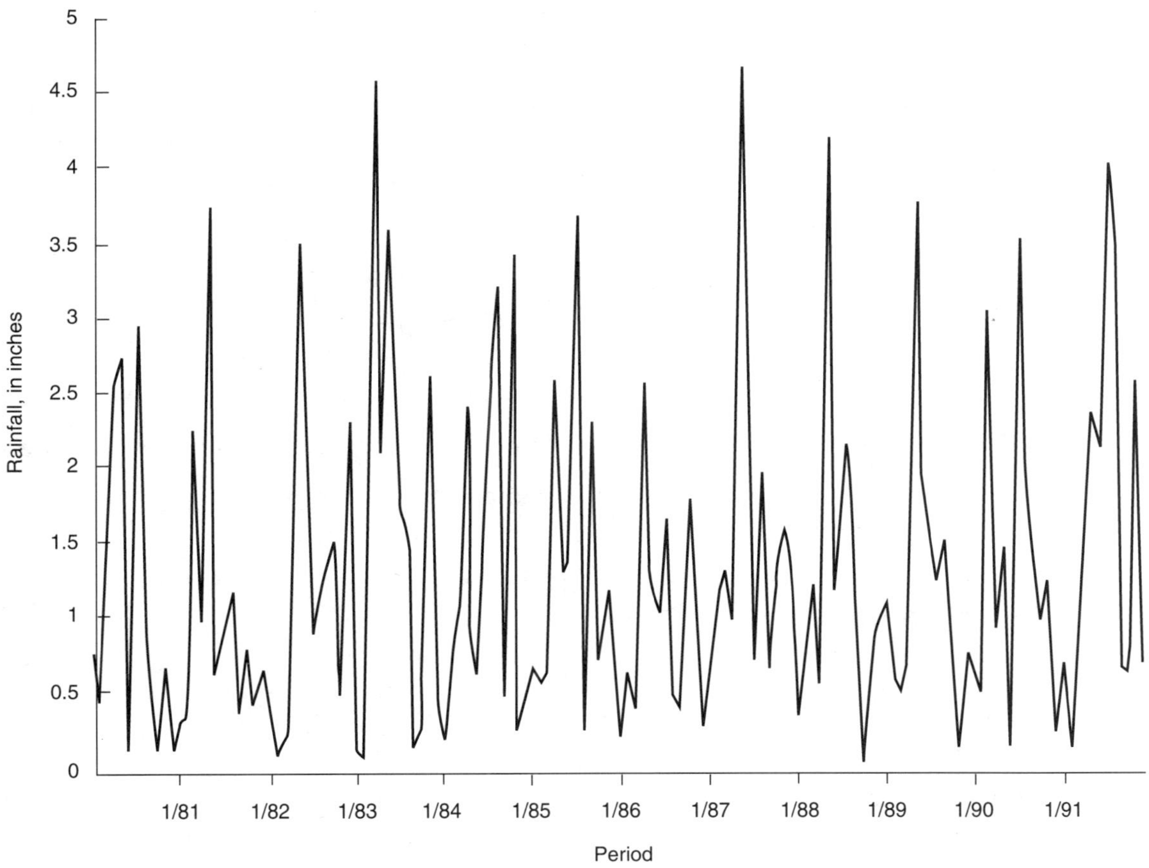

Figure 8-2 Monthly precipitation

to be positively related. The computed simple correlation between these two variables is 0.85. People use more water when the weather is hotter.

The correlation between water sales and precipitation is +0.25, suggesting that people use more water when it rains. This simple correlation, however, ignores the positive correlation between rain and temperature. Thus, the real reason people use more water during months of higher rainfall in this particular location is because these are also the hottest months.

One transformation that often is productive in this context is to express the weather variables as differences between their long-term averages and the sample observations. Thus, average monthly temperatures for July are averaged over a long period, ideally 30 to 40 years. Then, differences between this long-term average July temperature and the year-by-year sample observations for average July temperature are computed. The resulting differences are used as the temperature variables in regressions with monthly or seasonal dummy variables, as described below.

Expressing the temperature and precipitation values for this sample period in terms of departures from their long-term averages for each month corrects the simple correlation between rainfall and water use. The simple correlation between *DELRAIN*, the difference between average monthly rainfall for the sample months and their respective long-term averages, and *MWATER*, monthly residential water use, is negative.

Multivariate Models

The influences of temperature and precipitation variables on water use can be analyzed with a multiple regression model. In the model, variations of water use (the dependent variable) are explained by variations in several independent or explanatory variables. A regression using these data, which was presented in chapter 6 as Eq 6-9, is

$$MWATER = \beta_0 + \beta_1\,DELTEMP + \beta_2\,DELRAIN + \beta_3\,TREND + \beta_4\,JAN + \beta_5\,FEB + \beta_6\,MAR + \beta_7\,APR + \beta_8\,MAY + \beta_9\,JUN + \beta_{10}\,JUL + \beta_{11}\,AUG + \beta_{12}\,SEPT + \beta_{13}\,OCT + \beta_{14}\,NOV$$

Where:

MWATER = monthly residential systemwide water sales, in thous-gal units

DELTEMP = deviation of average monthly temperature, in °F, from long-term average monthly temperature

DELRAIN = deviation of total precipitation, in inches, from long-term average monthly precipitation

TREND = time index ranging from 1 to 132

JAN through *NOV* = dummy variable for each specified month, equal to 1 during that month and 0 otherwise

The individual estimated β coefficients in Table 8-1 show the effect of each of the independent variables on water use when all other influences represented in the equation are held constant.

The coefficient of 5,910 on *DELTEMP* means that an increase in average monthly temperature of 1°F can be expected to increase water use in this system by 5,910 thous gal. Note that the regression coefficient on *DELRAIN* is negative, as expected, although the *t* statistic of 0.91 is only significant at 0.18, meaning that there is slightly better than a four in five chance that people actually do use less water when it rains (and a one in five chance that they use more). The significance level is half of that shown as the *P* value in Table 8-1 since a one-tailed test is used. Thus, the analyst would not place much confidence in the value of the coefficient on rain.

The *t* statistic for *DELTEMP* is 2.12, meaning that there is only a 1 in 50 chance that the true effect of temperature is contrary to our expectations, and there is a high level of confidence that this relationship is accurately portrayed. When the equation is adjusted for first order autocorrelation[*] of the residuals, the estimated coefficient for rain becomes more significant.[2]

The regression accounts for 88 percent of the variation of water use from its mean value over the 132 observations used in estimating the model. At least some of the unexplained variation in water use may be due to peculiarities of individual household water use. Irrigation decisions based on the time of year would be effectively modeled by the monthly dummy variables, and those motivated by actual weather conditions by temperature

---

[*] Refer to chapter 3 for a discussion of autocorrelation.

Table 8-1 Summary output: Monthly aggregate residential water use equation

Regression Statistics	
Multiple *R*	0.94
R^2	0.89
Adjusted R^2	0.88
Standard Error	108,382.12
Observations	132

Analysis of Variance

	df	*SS*	*MS*	*F*	Significance
Regression	14	1.13769*E*+13	8.12639*E*+11	69.18	1.0189*e*–
Residual	117	1.37436*E*+12	11,746,683,368		
Total	131	1.27513*E*+13			

	Coefficients	Standard Error	*t* Stat	*P* Value	Lower 95%	Upper 95%
CONSTANT	72,952	37,278.02	1.96	0.05	-875.19	146,779.12
DELTEMP	5,910	2,782.38	2.12	0.04	399.73	11,420.45
DELRAIN	-10,579	11,584.46	-0.91	0.36	-33,521.71	12,363.10
TREND	2,859	249.14	11.48	0.00	2,365.95	3,352.76
JAN	-9,497	46,295.48	-0.21	0.84	-101,182.92	82,188.59
FEB	-17,968	46,281.40	-0.39	0.70	-109,626.03	73,689.72
MAR	6,976	46,268.65	0.15	0.88	-84,656.40	98,608.88
APR	181,274	46,257.25	3.92	0.00	89,663.49	272,883.60
MAY	360,452	46,247.18	7.79	0.00	268,862.12	452,042.35
JUN	636,623	46,238.46	13.77	0.00	545,050.18	728,195.86
JUL	739,715	46,231.08	16.00	0.00	648,156.29	831,272.72
AUG	605,796	46,225.03	13.11	0.00	514,249.78	697,342.28
SEPT	428,156	46,220.33	9.26	0.00	336,618.94	519,692.83
OCT	161,045	46,216.98	3.48	0.00	69,514.55	252,575.14
NOV	38,555	46,214.96	0.83	0.41	-52,971.70	130,080.91

NOTE: *P* values are for a two-tailed test.

and rainfall. However, landscape irrigation based on the appearance of the lawn or on levels of human comfort may differ markedly from the needs of plants. Many people with timer-controlled irrigation systems simply set the timers, and do not adjust them as the weather changes, except to turn off the system when the irrigation season is over. Automatic sprinkler systems frequently continue to operate even during a rain storm. Such human foibles are hard to model.

An additional source of potential error in an estimation model such as that previously described is that in many urban areas precipitation may be far from uniform. One side of town may be drenched in rain while the other side is dry. When a grid of rainfall data for the service area is unobtainable, the resulting errors in measuring rainfall result in commensurate errors in forecasting water demand.

Other Weather Variables

Numerous weather variables have been suggested and used as alternatives to simple temperature and precipitation in water demand studies.

Individual Weather Variables

The rainfall variable can be divided between the usual irrigation months and the remainder of the year. In some studies, only rainfall over a threshold, such as 0.1 in., is counted, or the number of significant days of rain per month is used.[3] This latter refinement might prove to be beneficial, since three rain days, each delivering ½ in. of precipitation, may reduce the need to irrigate more than one storm delivering 1½ in. The effect of temperature or rainfall may also be nonlinear, so that a better fit would be obtained with a two-part variable, such as temperature above and below 86°F. Additional weather variables that might be used include humidity, wind speed, daylight hours, and pan or pond evaporation.

Potential or Measured Evapotranspiration

Evapotranspiration is the combined water loss from the soil and the leaf surfaces of plants and is a key element of the hydrologic cycle. Since evapotranspiration is partially controlled by vegetation and partially by factors related to soil type, it is difficult to measure. Given the limited availability of field measurements, many researchers focus on a theoretical value referred to as potential evapotranspiration. This value can be computed using climatic and weather variables available from the National Weather Service. *Potential evapotranspiration* is the water loss from a vegetation cover that never suffers from a lack of water and is primarily a function of the energy available for evaporation.[4] Thornthwaite and Mather (1955) fitted evaporation data from watersheds and irrigation plots to air temperature, obtaining the following expression for unadjusted potential evapotranspiration in centimetres per month:

$$E = 1.6\left(\frac{10t}{I}\right)^{a} \tag{8-1}$$

Where:

t = average monthly temperature, in °C

I = is an annual heat index determined from summing the 12 monthly heat index values

a = a function of the heat index as indicated in Eq 8-2.

$$a = \frac{I^3}{6.75(10^7)} - \frac{I^2}{7.71\,(10^5)} + \frac{I}{1.79\,(10^2)} + 0.49 \tag{8-2}$$

The classic Howe and Linaweaver (1967) research conducted in the 1960s on a nationwide sample of residential customers employed the Thornthwaite method of estimating evapotranspiration.

Some research suggests simpler weather variables perform as well or better than potential evapotranspiration.[5] Direct field measurements of evapotranspiration are sometimes available, and previous measured values can be adjusted for actual temperature and rainfall after-the-fact to provide an effective way to model the weather.[6]

Cycles

One way of looking at weather cycles is by tallying "heat waves." *The Weather Almanac* (Ruffner 1987) explains

> There is nothing in American climatological annals to touch the heat waves which came with the Dust Bowl droughts of the 1930s. The years 1930, 1934, and 1936 brought progressively more severe summer weather. Record highs of 121 degrees in North Dakota and Kansas, and 120 degrees in South Dakota, Oklahoma, Arkansas, and Texas were observed in the ugly summer of 1936. July and August of that year saw record highs of 109 degrees or better tied or broken in Indiana, Louisiana, Maryland, Michigan, Minnesota, Nebraska, New Jersey, Pennsylvania, West Virginia, and Wisconsin .... The heat wave of July 1966 covered much of the eastern and middle continent with high temperatures and very high humidity .... The highest death toll for a year since the 1930s belongs to 1980, according to the National Center for Health Statistics. That year saw unusually high temperatures hold the central and southwestern United States in their grip for the best part of 15 weeks .... The 1980 heat wave ... began about June 10, and except for remissions from about July 21 to August 10, and August 17 to September 1, lasted continuously through September's third week. As an example of its intensity, the first week of September saw temperatures of 9 degrees above normal over much of the central and eastern states.

North American weather patterns depend a great deal on the position and orientation of the jet stream. As winter abates and the continent warms, the jet stream usually shifts northward, moving thundershowers to the plains and humidity to the Midwest. A "dust bowl pattern" exists if the prevailing winds from the Southwest sweep farther north than usual in a more or less self-perpetuating circulation pattern.

It appears that heat waves and drought events are periodic. The climatologist J. Murray Mitchell conducted a statistical study of tree rings and found a 20- to 22-year drought cycle, persistent since 1700 (Schneider and Londer 1984).

Local exceptions usually exist with respect to area-wide heat waves and drought events. Studies of the cycle of extremely dry, hot weather for specific localities often suggest the existence of a tendency for recurrence at a more or less fixed cycle. Such tendencies, however, usually explain only a fraction of the drought events and predict drought periods when none have occurred in the historic record (Kane and Teixeira 1991).

Weather cycles, therefore, do not appear to be sufficiently predictable for inclusion in long-term water demand forecasts. However, the historic record of temperature and precipitation should be consulted to determine the likely range of variability of these key factors in influencing water use. Hypotheses about weather cycles might be used in forecasts prepared in connection with drought contingency plans.

Trends

Climatic trends such as heat-island effects and global warming offer potential effects for inclusion in long-run urban water demand forecasts. The issue is whether these factors are understood well enough to be used.

Heat-Island Effects

Heat-island effects are increases in average temperature throughout the year in the microclimate of larger metropolitan regions. This phenomenon is well documented and could be included in long-term water demand forecasts.

Temperature trends can be identified in many major urban areas (see Cayan and Douglas 1984; Maidment and Djokic 1990). Many areas of Los Angeles, Calif., for example, provide evidence of heat-island effects, with annual temperatures rising on average 0.5°F per decade, as suggested by Figure 8-3. Each temperature series exhibits a statistically significant and positive slope in a simple time-trend regression. Heat-island effects appear to be the primary contributor to this local trend.

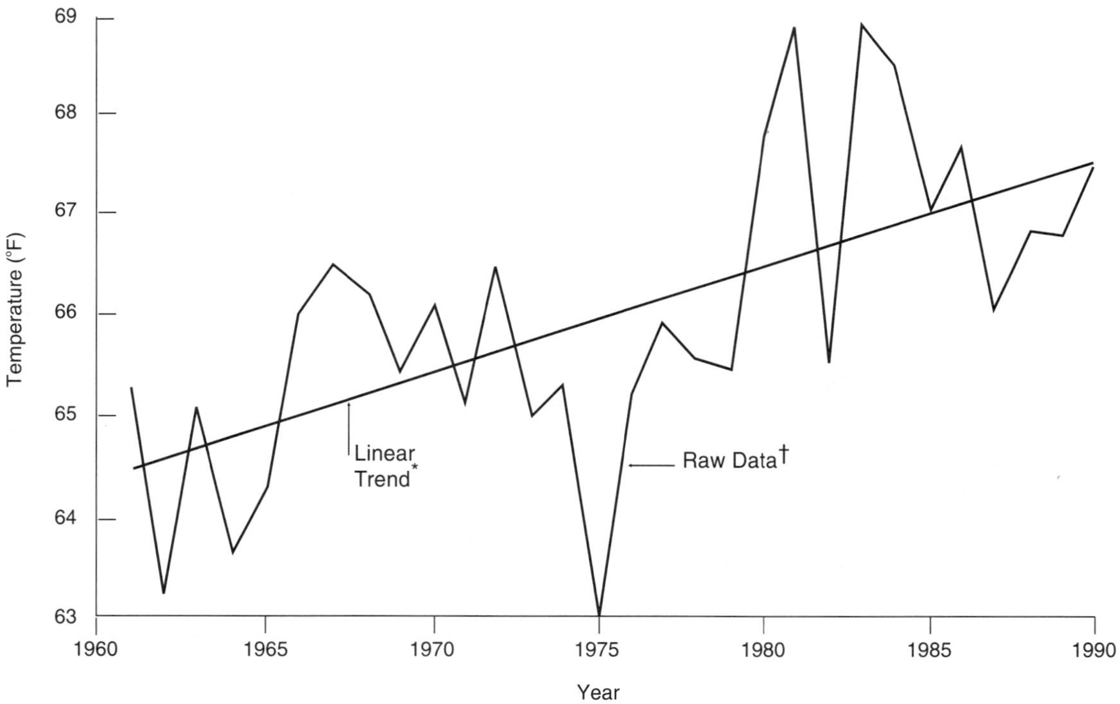

Linear trend computed by regression.
†*Raw temperature data taken from National Oceanic and Atmospheric Administration (1991).*

Figure 8-3 Annual average temperature: Los Angeles, Calif., Civic Center

Because temperature trends resulting from heat-island effects are statistically significant, they appear to be suitable for inclusion in long-term forecasts. Their use requires a prediction of the urban growth rate. Since a growth rate forecast of population, commercial development, and industrial location is central to every long-term water demand forecast, the addition of heat-island effects does not add significant complexity to the forecasting task and may provide a small increase in forecast accuracy.

Global Warming

Predictions of global warming are fraught with great uncertainty and are therefore of little value to the urban water demand forecaster. If it occurs, however, some of the most dramatic effects of global warming on water demand will be via the number of extremely hot days. A 2°F hike in average temperature results in a substantial increase in the probability of long runs of hot days.

Summary

Climate and weather are the primary causes of seasonal patterns of urban water use and the day-to-day fluctuations in use levels. Average monthly temperature and precipitation often track fairly closely with the monthly pattern of urban water use, as the example developed in this chapter illustrates. Multiple regression is one way to relate temperature and precipitation to system water sales to characterize and predict aspects of the seasonal pattern, such as the monthly peak-to-average ratio of water demand. Evapotranspiration is often used as an alternative weather variable.

Long-term forecasts of drought cycles and climatic change are subject to considerable uncertainty. Urban heat-island effects, however, are well established.

Weather and climate can explain much of the volatility of actual water use, thereby providing an effective avenue for exploring this aspect of urban water demand.

Notes

1 The correlation coefficient between x and y, denoted as r, is

$$r(x, y) = \frac{\left[\sum_i (y_i - \bar{y})(x_i - \bar{x})\right]}{\sqrt{\left[\sum_i (y_i - \bar{y})\right]\left[\sum_i (x_i - \bar{x})\right]}}$$

2 There are several procedures for correcting first and second order autocorrelation that are generally available in an econometric package. For a discussion of these methods, refer to an econometrics book, such as *Econometrics* (1977) by G.S. Maddala. Available from McGraw-Hill, New York.

3 An interesting class of time-series models incorporates "state dependent" effects (see Miaou, Shaw-Pin 1990. A Class of Time Series Urban Water Demand Models with

Nonlinear Climatic Effects. *Water Resources Research*, 26(2):169–178). State dependent effects occur when, for example, the effect of a 1° F increase in average (daily, weekly, monthly, or annual) temperature is not assumed to be independent of other climatic and socioeconomic conditions at the time it is experienced. Advanced estimation techniques requiring numerical optimization and maximum likelihood usually are required. The Miaou application concludes that the number of days of rainfall per month is a significant weather variable.

[4] Water-filled pans, such as the US Weather Bureau class A evaporation pan, may provide only approximate measures of evapotranspiration. For a detailed discussion of the measurement problem and other issues related to evapotranspiration, see Mather, J.R. 1984. *Water Resources: Distribution, Use, and Management*. New York: John Wiley & Sons.

[5] Using data for 33 cities in Southern California for 1970, Douglas and Smolen (1976) compared the performance of potential evapotranspiration with average monthly temperature and precipitation.

[6] A normalized monthly evapotranspiration series, adjusted for actual temperature and rainfall, was used by Billings and Agthe (1980).

References

Billings, R.B., and Agthe, D.E. 1980. Price Elasticities for Water: A Case of Increasing Block Rates. *Land Economics*, 56(2):73–84.

Cayan, D.R., and A.V. Douglas. 1984. Urban Influences on Surface Temperatures in the Southwestern United States During Recent Decades. *Journal of Climate and Applied Meteorology*, 23(11):1520–1530.

Douglas, M.W., and J.C. Smolen. 1976. Climatic Indicators in the Estimation of Municipal Water Demand. *Water Resources Bulletin*, 12(3):511–518.

Howe, C.W., and F.P. Linaweaver Jr. 1967. The Impact of Price on Residential Water Demand and Its Relation to System Design and Price Structure. *Water Resources Research*, 3(1):13–32.

Kane, R.P., and N.R. Teixeira. 1991. Power Spectrum Analysis of the Annual Rainfall Series for Massachusetts (NE, USA). *Climate Change*, 18(1):89–94.

Maidment, D.R., and D. Djokic. 1990. Weather-Related Water Demand Variability in Metropolitan Water District Service Area. Report submitted to the Metropolitan Water District of Southern California.

National Oceanic and Atmospheric Administration. 1991. *Local Climatological Data, Annual Summary for 1990, Los Angeles, Calif.* Ashville, N.C.: NOAA.

National Weather Service. 1988. *A Result for Temperature and Precipitation*. NOAA Technical Memorandum, NWS FCST 31, August 1988.

Ruffner, J.A., and F.E. Bair, editors. 1987. *The Weather Almanac: A Reference Guide to Weather, Climate, and Air Quality in the United States and Key Cities*, 5th ed. Detroit, Mich.: Gale Research Company.

Schneider, S.H., and R. Londer. 1984. *The Coevolution of Climate and Life*. San Francisco: Sierra Club Books.

Thornthwaite, C.W., and J.F. Mather. 1955. *The Water Balance, Publications in Climatology*. South Centerton, N.J.: Drexel Institute of Technology, Laboratory of Climatology.

Wagner, J.A. 1989. Medium and Long-Range Forecasting. *Weather and Forecasting*, 4(3):413–426.

Chapter 9

Price Effects*

Evidence shows that water customers cut back their water purchases when water rates increase, with impacts noted in the short run (a few weeks or months) and long run (multiple years). Over short time periods, customers first reduce water uses they value least, such as washing sidewalks, buildings, and cars, and, to an extent, irrigating landscape. Over longer time periods, customers may replace water-using appliances, devices, and equipment with more water-efficient models; adopt low-water-use landscaping innovations; and make other investments that result in increased efficiency in water use.

These effects often are subtle, given the typically low increases in water rates, and are assured only if water rates keep pace with inflation. However, customer responses to water rates and fees translate directly to the bottom line—water utility revenues. It is important, then, to strive for accuracy in projecting the effect of changes in water rates.

This chapter presents an overview of common types of water rate and fee schedules, and an explanation with examples of price and revenue elasticities. The discussion includes: (1) impacts of water metering, (2) controlled comparisons of communities with similar socioeconomic and weather conditions, but different levels of water rates, and (3) econometric studies of water demand. The chapter shows how "first-cut" assessments of the impact of water rate changes can be developed with price elasticities from published research.

Schedules of Rates and Fees

Many types of rate schedules and fee arrangements exist in the potable water industry, including the following:

- a monthly service fee plus a constant volume charge
- a flat rate
- a decreasing block rate schedule

*Donald E. Agthe co-authored this chapter.

- an increasing block rate schedule
- a minimum use included with the service charge
- emergency drought rates to ration shortages
- seasonal rates
- "lifeline" rates for elderly and low-income groups, often featuring significant discounts for initial, basic necessity quantities of water

Monthly Service Fee and Constant Volume Charge

A typical rate schedule combines a monthly service fee for each connection with a constant volume charge for the amount of water used. This is graphed in Figure 9-1, in which the solid, dark bar situated at 0 quantity on the price axis signifies a $5.00 fixed fee per billing period.

Decreasing Block Rates

Decreasing block rate charges, such as the two-part rate schedule in Figure 9-2, are widespread. Initially known as the "Hopkinson" rate schedule in electric power circles, the decreasing or declining block rate schedule has lower block rate charges for higher quantities of water usage or purchase. In Figure 9-2, for example, a rate of $3.00/thous gal is charged for the first 3,000 gal of water usage per billing period. Purchases of quantities of water in excess of 3,000 gal and up to 6,000 gal in a billing period are charged at $2.50/thous gal. Finally, water purchases in excess of 6,000 gal a billing period are

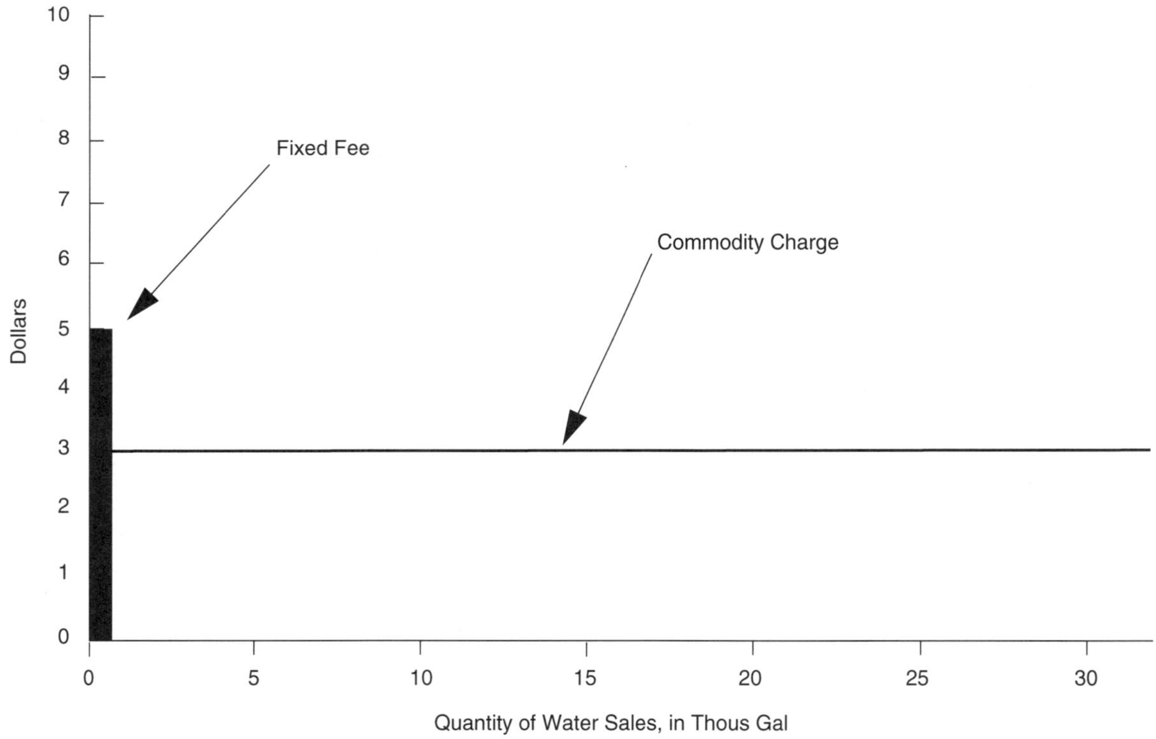

Figure 9-1 Fixed fee and constant commodity charge

priced at $2.00/thous gal. Again, the $5.00 fixed fee for the service hookup is indicated by a solid, dark bar situated along the price axis.

The decreasing block rate schedule is often considered to be a marketing tool, i.e., a tactic to increase water sales. Other functions of the decreasing block rate schedule include the ability to specify a rate schedule that will charge lower rates to industrial water users than to residential or smaller commercial customers. This objective can also be achieved by separate industrial and commercial rate schedules.

Increasing Block Rates

Currently, many utilities are moving to increasing block rates to provide incentives for larger-volume customers to conserve water. As its name suggests, an increasing block rate schedule reverses the pattern of a decreasing block rate schedule, charging higher rates for greater quantities of water consumption. Figure 9-3 shows an increasing block rate schedule with the same block rate boundaries as the decreasing block rate schedule of Figure 9-2. Table 9-1 presents these three rate schedules as lists of rates and fees.

Table 9-1 Typical block rate schedules

Usage	Rates per Thous Gal		
	Constant	Declining	Increasing
First 3,000 gal	$3.00	$3.00	$2.20
Second 3,000 gal	$3.00	$2.50	$2.60
All remaining water	$3.00	$2.00	$3.20
Monthly service charge: $5.00			

The rate in the final block of an increasing block rate structure could be very high (punitive) if the need is to curtail water use because of a drought or a long-term supply limitation. Such punitive rates often are imposed as part of a crash conservation program. One Utah system charged a penalty rate of $10.00/thous gal for usage exceeding the mandatory limit and experienced a 50 percent overall drop in water use. Some California utilities experienced reductions of 80 to 95 percent in their top use block in response to high penalty rates.

What Is the "Price" of Water?

When determining the price of water, the rate schedules and fee arrangements previously listed can be divided into two categories: unmetered service and metered service.

Unmetered Service

A constant bill regardless of how much water is purchased by a customer is the so-called "flat rate" charge used in areas without water meters. Flat rate charges provide no incentive to conserve or economize on water use, and, as noted subsequently in this chapter, typically result in higher levels of water usage. In this case, the price of additional

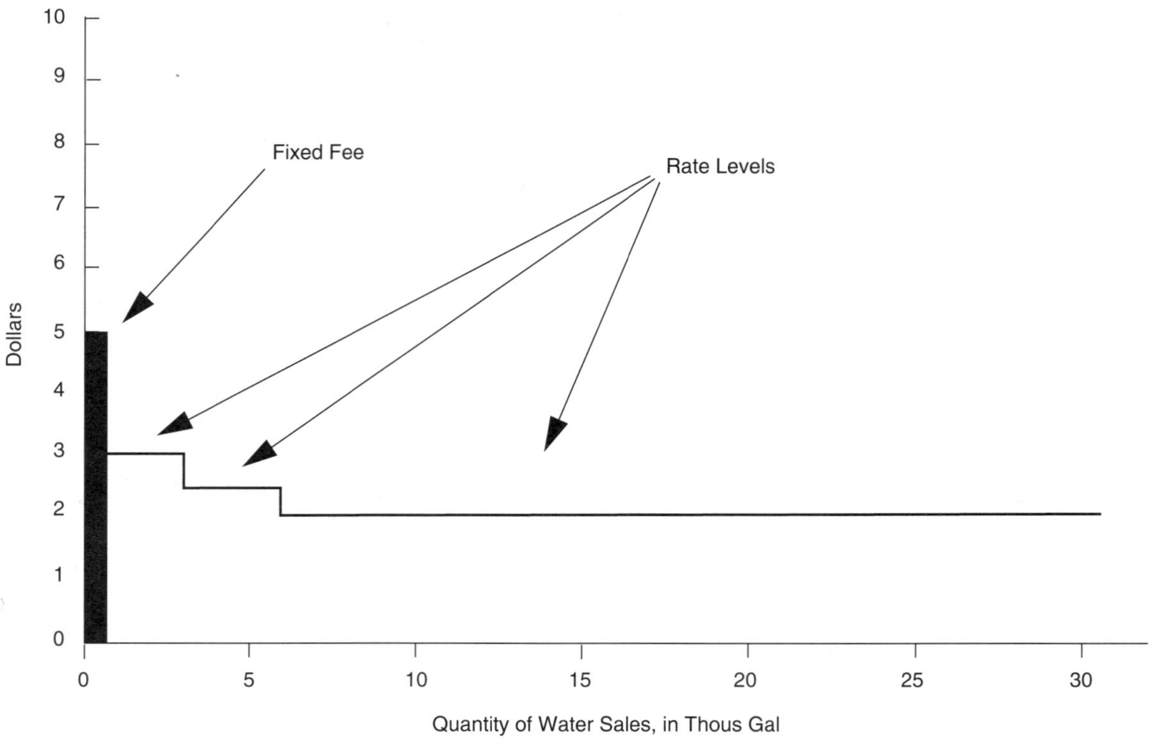

Figure 9-2 Decreasing block rate schedule with fixed fee

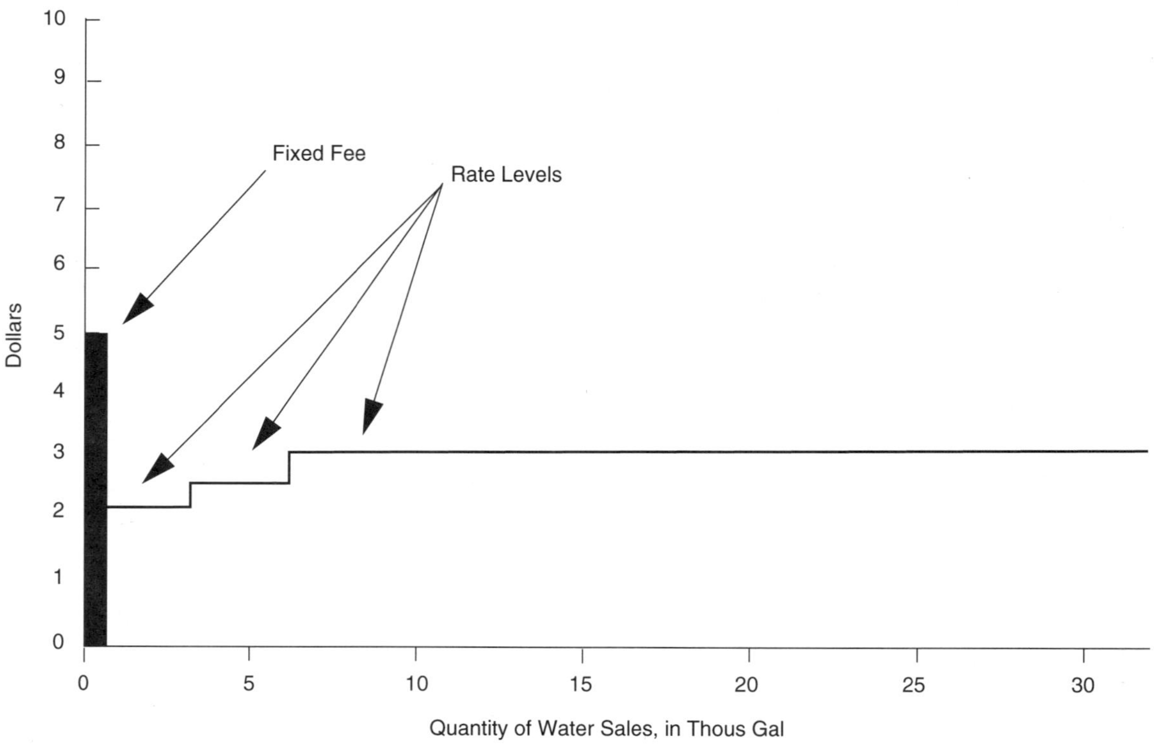

Figure 9-3 Increasing block rate schedule with fixed fee

quantities of water to an individual customer is always 0, although there is a cost of obtaining service.

Metered Service With Commodity Charges

If the water bill contains components that vary with the amount of water purchased—a commodity charge—consumers are likely to react to increases in their water bill by reducing their water purchases. Then it becomes necessary through a regression or other statistical analysis to estimate consumer price effects.

This requires translating the schedule of rates and charges applying to water sales to a number or numbers representing the price of water. This is the issue of the *price specification*, which figures importantly in many published studies of urban water price impacts. It must be determined which rate or combination of rates triggers and is quantitatively linked with the consumer's purchasing decisions.

The literature is largely concerned with the following major price specifications, described in note 1 at the end of this chapter:[1]

- average price
- marginal price
- marginal price plus "difference"
- total bill

Fortunately for the utility analyst preparing a forecast of water use, the results of prior studies, including the estimated price elasticities of demand, can generally be used regardless of which price specification was used to derive them. The choice of a price specification only becomes critically important when rate structures are being changed. If consumers respond primarily to their total water bill rather than to marginal price, a revenue-neutral modification of the rate structure will have little or no effect on aggregate demand and can be ignored for forecasting purposes. The imposition of a very high, or punitive, marginal price for usage above some selected quantity, however, will have a dramatic impact on demand that cannot be ignored. In this situation, water customers should be disaggregated by volume use categories in order to intensively analyze those who will be subjected to the punitive rates.

For proportional changes in all elements of a rate structure (e.g., monthly service charges and marginal prices), similar effects on water usage and revenue are predicted by all of the price specifications. For such rate increases, therefore, it is possible to rely on price elasticity estimates obtained from diverse price specifications.

Price Elasticities of Demand and Revenue

The *price elasticity of demand* measures the responsiveness of the quantity demanded to a change in price. It is computed as the percentage change in quantity of water sales or usage divided by the percentage change in price/rates, when other factors, such as income and weather, are constant or controlled in the estimating equation. It is often defined as the percentage change in water purchases associated with a 1 percent change in water price.[2] The price elasticity is a pure number, independent of the dimensions in which water sales and prices are measured.

As long as the demand curve for water has a negative slope (i.e., people buy less water when the price rises), the price elasticity of demand will be numerically negative (less than 0). While some individuals may ignore a price increase and continue with their usual purchases, it is highly unlikely that anyone would respond to a price increase for water by buying more water.

The *price elasticity of revenue* shows the responsiveness of total revenue to a change in price. It is defined as the percentage change in total revenue divided by the percentage change in price. The price elasticity of total revenue is always equal to 1 plus the price elasticity of demand. Thus, the price elasticity of revenue may be either positive or negative.

Measured price elasticities of demand between 0 and –1 are referred to as relatively inelastic. That is, quantity is relatively unresponsive to a change in price since the percentage change in quantity is less than the percentage change in price. The corresponding price elasticity of revenue is between 1 and 0.

For example, given a water utility that is considering raising water rates by 15 percent, and an estimate of the systemwide price elasticity of demand of –0.3, the effect on both water sales and system revenue can be directly computed. The projected price increase will reduce water sales by 4.5 percent (the elasticity of –0.3 times the price increase of 15 percent).

The price elasticity of total revenue in this example is 0.7 (1 + –0.3). Revenue is expected to increase by 10.5 percent (+0.7 × 15 percent) when price is increased by 15 percent, assuming nothing else changes. Thus, while water rates increased by 15 percent, revenues increased by only 10.5 percent because water customers responded to the price increase by reducing water use by 4.5 percent.

A price elasticity of demand exactly equal to 0 is known as *perfectly inelastic* and indicates that no change in water sales takes place in response to a change in price. A perfectly inelastic demand results in a revenue elasticity of 1; total revenue increases by the same proportion as price. While some individuals may have a perfectly inelastic demand for water, it is highly unlikely that this would be true for an entire community.

A price elasticity of demand less than –1 indicates that the percentage change in quantity is greater than the percentage change in price, a situation known as *relatively elastic*. A price increase when demand is relatively elastic results in less revenue. For example, the revenue elasticity corresponding to a price elasticity of –1.4 is –0.4 (1 plus the price elasticity of demand). In this event, a 15 percent increase in water price would cut sales by 21 percent and revenue by 6 percent. Fortunately for the water utility industry, the price elasticity of demand for water is usually inelastic, so that the price elasticity of revenue is positive.

A price elasticity of demand of exactly –1 is called *unitary elastic*. It means that the percentage changes in price and quantity are equal. The corresponding revenue elasticity is exactly 0.

Price elasticities are usually "local" estimates, applicable only within a range around the price where they are initially estimated. For the linear demand curve (Eq 9-1), for example, the price elasticity for a price/quantity combination (P^*, Q^*) is $b_1(P^*/Q^*)$, where $b_1 < 0$. At low prices, this is inelastic, while at a high enough price, this will become price elastic. This suggests that further rate increases will result in a decrease in overall revenues.

This relationship, i.e., price inelastic responses at low price and high quantity, and price elastic responses at high price and low quantity, is true of all linear demand curves.

Price elasticities are much like approximations found in physics. They provide insight into the implications of rate changes, but require skill in their manipulation.

It is important to note that the estimates using elasticities are sensitive to inflation. Consumers typically take inflation into account when they decide what and how much to buy. After 1 year, if inflation is running at 4 percent, a 15 percent water rate increase deteriorates to 11 percent in terms of its impact on water purchases. After only 4 years of 4 percent price inflation, the impact of the initial 15 percent rate increase erodes to a price reduction of 2 percent in real terms (with compounding). This dissipation of rate effects is quicker for more rapid inflation. When rate increases are smaller than the general rate of inflation, water prices drop in real terms.

Impact of Metering

Statistical studies and practical experience on the part of utility officials have shown that water use decreases when water meters are installed and commodity charges are imposed on water customers. Typical reductions in water use range from 15 to 40 percent when meters are installed.[3]

A rigorous comparison of metered and unmetered water use was developed by the Johns Hopkins Residential Water Use Project, funded by the Federal Home Administration. A sample of 39 study areas in the United States, each including a relatively homogenous residential district, were master metered during the period 1961 to 1965. Flows of water were recorded every 15 minutes. The study included 10 metered areas with public sewers in Oakland, Los Angeles, and San Diego, Calif.; 5 metered areas with septic tanks in Des Moines, Iowa, Great Falls, Mont., and Denver, Colo.; and 5 apartment building areas without individual residential meters in San Diego, Denver, and Washington, D.C. Linaweaver, Beebe, and Skrivan (1966) reported dramatically lower water use per household in areas with meters and commodity charges (shown in Table 9-2).

Table 9-2 Water use in metered and flat rate areas, western United States

	Gal/Day/Dwelling Unit	
Annual Average	Metered	Flat Rate
Leakage and Waste	25	36
Interior Use	247	236
Sprinkling	186	420
Total Household Use	458	692
Peak Hour	2,481	5,170

Source: Linaweaver, Beebe, and Skrivan (1966).

Controlled Comparisons

While it may not be easy to identify exactly what the relevant price of water is, water customers do react to variations in water prices when there is a relationship between the amount of water purchased in a billing period and the total water bill.

Table 9-3 presents a controlled comparison of average water use per metered single-family residence in two nearby Arizona communities. These communities have approximately the same household income levels, household sizes, and other characteristics. Rainfall and temperature were similar in both areas, and account for most shared year-to-year variation. Over the 6 years recorded, the average price of water in one district was 65 percent higher than in the adjoining area, while average usage was 28 percent lower.

Table 9-3 Controlled comparison of water usage rates

	Average Monthly Bill and Water Sales Per Customer			
	Tucson Water		Flowing Wells	
Year	Water Bill	Sales, *thous gal*	Water Bill	Sales, *thous gal*
1975	$6.50	12.3	$3.90	16.5
1976	9.32	10.7	6.31	15.2
1977	11.04	9.5	6.37	13.7
1978	11.71	9.4	6.79	12.8
1979	11.61	9.2	7.06	12.6
1980	$11.20	9.6	$6.66	13.0

Source: Billings and Day (1982).

NOTE: The water bills used to measure the price of water were computed as the average of the bills for 8,976 gal/month using winter rates and 14,960 gal/month using summer rates. Bills are stated in 1974 dollars.

Econometric Studies

The third and most comprehensive evidence for the existence and size of customer responses to water rates comes from an extensive collection of econometric studies. These typically estimate a multivariate regression, such as the simple linear relationship

$$Q = b_0 + b_1 P + b_2 X_2 + \ldots + b_k X_k \qquad (9\text{-}1)$$

Where:

Q = water use

b_0 = constant or intercept of estimated regression

$b_1, b_2 \ldots b_k$ = coefficients estimated by the regression, which show the effect of each independent variable on Q

P = the real or inflation-adjusted price of water

X_i, $i = 2,3,\ldots,k$ = other explanatory variables, such as personal income, household size, rainfall, and temperature[4]

Usually, research results pertaining to the "price elasticity of demand" are provided, in addition to estimated regression coefficients and diagnostic statistics.[5]

A review of more than 40 statistical studies of urban water demand and its components suggests quantitative and qualitative facts about water demand price elasticities.

Qualitative Findings

- Demand is more elastic (or less inelastic) for outdoor (sprinkling) than for indoor use.
- The response to price changes increases with time, as long as these changes are not eroded by inflation. There may be a temporary "shock effect" of a price increase resulting from water users overestimating the impact of the increase on their water bills. Typically the adjustment to much higher (or lower) water prices will occur over an extended period of time as fixtures and landscaping are altered.
- Price elasticity increases with the ratio of water bills to income. This can be expressed in the following three corollaries:

  1. Demand is more elastic at higher prices. This is because at higher prices water bills take a larger proportion of income.
  2. Demand is more elastic for higher-volume uses. Within any well-defined category of customers, those who use larger volumes of water, other factors being equal, will find water bills more burdensome than their counterparts. It is also likely that those using a larger volume of water have adopted marginal uses for water that are avoided by those with smaller usage levels. A price increase tends to encourage these users to reduce or abandon some of these "extra" uses of water, such as hosing-down sidewalks or forgetting to turn off faucets.

     Commercial and industrial demand also tends to be more elastic for those who use larger than average volumes. The demand for water by an electric generating plant using a larger volume of water than the industry average will have a higher (long-run) elasticity. Their short-run elasticity may be near 0 if reducing water use requires new capital investment, but given time for adjustment, the larger outlays for water will provide an incentive for water-saving changes in production methods.
  3. Demand becomes less elastic as incomes rise. With higher incomes, water bills take a smaller proportion of the household budget, making customers less sensitive to its price.

Quantitative Findings

Residential Price Elasticity of Water Demand

Excellent quantitative estimates of water demand price elasticities are available in a study based on data from 218 US cities (Foster and Beattie 1979). This research calculates variations in water use per household as a function of (1) average price, (2) median household income, (3) rainfall during the growing season, and (4) average number of residents per household. The study uses a functional form that makes the estimated price elasticity proportional to the price.

Table 9-4 presents these price elasticity estimates from six geographic regions of the nation for three price levels, $1.00, $3.00, and $5.00/thous gal in 1990 dollars. The first column presents the price coefficient estimated in this research, while the remaining columns show the variation in price elasticity as the price of water increases. At $5.00/thous gal in 1990 prices, water demand in the Plains states and Rocky Mountain region, as well as in the Northwest, becomes price elastic. This implies, among other things, that utility revenues could be expected to drop with further rate increases.

The Foster and Beattie results can be compared with findings of selected price elasticity studies, shown in Table 9-5. These price elasticity values for annual, winter, summer, and outdoor sprinkling by city or region are presented along with the prevailing price of water, adjusted to 1990 dollars. These results indicate a clear tendency for the elasticity to be higher when the price is higher, thus supporting the functional form used by Foster and Beattie. Sprinkling demand is always more elastic than indoor water use, and equivalently, summer use is more elastic than winter use, since most winter use is indoors.

Table 9-4 Estimated price elasticities by region for selected price levels

Region	Coefficient, in *1990 dollars*	Elasticities and Selected Prices		
Price per Thous Gal		$1.00	$3.00	$5.00
Northeast	-0.20	-0.20	-0.60	-1.00
Midwest	-0.14	-0.14	-0.41	-0.68
South	-0.16	-0.16	-0.47	-0.79
Plains, Rockies	-0.38	-0.38	-1.15	-1.92
Southwest	-0.21	-0.21	-0.62	-1.04
Northwest	-0.46	-0.46	-1.37	-2.28

Source: Foster and Beattie (1979).

Industrial Water Price Elasticities

Table 9-6 presents results from three studies of the industrial demand for water. These results indicate industrial water use decisions are also related to the price of water. The range of price elasticities for the chemical industry reflects varying price and production conditions faced by firms in different regions. The range of values reported also suggests that utilities serving large industrial demands for water need to carry out well-designed studies of industrial demand in order to accurately include these responses in their forecasts.

Commercial Water Price Elasticities

Most commercial accounts use water in the same way as households, i.e., indoor water use in bathrooms or for drinking and outdoor water use for landscape irrigation. When residential and commercial accounts have no irrigated landscaping, demand tends to be quite inelastic (between 0 and -0.3). Commercial accounts with irrigated landscapes exhibit higher elasticities, perhaps ranging up to -0.7.

Table 9-5 Estimated price elasticities, results of demand studies

Area	Annual	Winter	Summer	Outdoor Sprinkling	Price, *1990 dollars*
Eastern US		-0.15	-0.57	-1.57	1.68
Western US		-0.15	-0.43	-0.70	1.51
Toronto, Ont.		-0.75	-1.07		2.22
Chicago, Ill.	0				0.96
Chicago suburbs	-0.52				3.27
Cincinnati, Ohio	-0.62				2.19
Kentucky	-0.92				6.79
Raleigh, N.C.		-0.31	-1.38		2.39
Miami, Fla.	-0.51				0.79
Kansas	-0.69				2.16
Texas		-0.19	-0.37		2.73
Utah	-0.77				0.98
Tucson, Ariz.	-0.52				0.98
Oakland, Calif.	-0.15	-0.10	-0.17		0.77
Honolulu, Hawaii	-0.35				0.51

Sources: (in order listed in the table) Howe (1982); Howe and Linaweaver (1967); Grima (1973); Wong (1972); Clark and Goddard (1977); Grunewald, Haan, Debertin and Carey (1976); Danielson (1979); Gibbs (1978); Gottlieb (1963); Griffin and Chang (1990); Gardner and Schick (1964); Billings and Agthe (1980); Weber (1989); and Muncur (1987).

Table 9-6 Industrial water demand elasticities

Industry	Elasticity
Chemicals	-0.74, -0.66
Metal Fabrication	-1.00
Stone, Clay	-0.38
Forest Industry	-0.51
Petrochemical	-0.12
Heavy Industry	-0.25
Light Industry	-0.54

Sources: Babin, Willis, Allen, and Vlachou (1981); DeRooy (1974); and Rensetti (1988).

Using Price Elasticities in Demand Forecasts

Price elasticities from preexisting studies can be used to estimate the impact on volume sales and revenues of a previously enacted rate change. They also can be used to determine the rate change needed to achieve a given revenue requirement. These computations are summarized by three simple formulas, and illustrated by the following examples.

$$\Delta Q = \frac{\Delta P}{P} \times E_p \times Q \tag{9-2}$$

$$\Delta R = \frac{\Delta P}{P} \times E_r \times R \tag{9-3}$$

$$\frac{\Delta P}{P} = \frac{\Delta R}{R} \times \frac{1}{E_r} \tag{9-4}$$

Where:

ΔQ = change in quantity

$\frac{\Delta P}{P}$ = proportional change in price or rates

E_p = price elasticity of demand

Q = original quantity

ΔR = change in revenue

E_r = price elasticity of revenue = $E_p + 1$

R = original revenue

Example Computation

A utility in Kentucky currently delivers 40 mil gal per day (mgd) at an average price of $2.00/thous gal, yielding revenues of $80,000/day. This water agency wishes to increase revenues by 8 percent next year to catch up with past inflation in the costs of supplies and labor.

Translating the Revenue Objective Into a Rate Increase

Most rate and revenue computations begin with an unconstrained demand forecast. This forecast is unconstrained in the sense that it ignores behavioral responses of consumers to higher rates. In essence, it assumes that the price elasticity of water demand is 0.

In this case, the unconstrained forecast suggests an 8 percent increase in utility rates and charges is needed to supply an 8 percent increase in utility revenues next year.

This should be understood to mean an 8 percent *real* increase in water rates and charges. If the objective is to close the gap between labor and supply costs and revenues, next year's inflation must be taken into account in determining the dollar amount of the rate increase. At minimum, in nominal dollar terms, the rate increase supplying 8 percent more in purchasing power to the utility next year equals 8 percent plus next year's anticipated rate of inflation. Utilities generally take this into account by projecting next year's budget expenditures, which include expected wage rate and materials price increases.

If 5 percent inflation is anticipated next year, rates and charges must rise by at least 13 percent to achieve an 8 percent real increase in revenue next year. When both 5 percent inflation and the most likely consumer responses to the price increase are taken into account, the required increase in water rates is close to 18 percent. The required computations will be presented after the discussion of selecting a price elasticity.

Selecting the Price Elasticities

Price elasticities for use in a forecasting model can be selected from Table 9-4 or Table 9-5, based on regional proximity or other similarities. An elasticity for the specific utility could also be estimated from a structural demand model (see chapter 3).

The coefficients from Table 9-4 can be used by choosing the appropriate regional value from the table and multiplying it by the prevailing price of water (in 1990 prices). The values from Table 9-4 are, in many cases, similar to those shown in Table 9-5 for specific locations.

Alternatively, a location similar in climate to the forecast area could be selected from Table 9-5.

In this example, Tables 9-4 and 9-5 provide as many as three possible elasticity coefficients for this Kentucky water utility.

Using the price elasticity for the South from Table 9-4 suggests a price elasticity of -0.32, which is the coefficient of -0.16×2 (the local price).

The price elasticity for the eastern United States in Table 9-5 could be used, taking the proportion of water used indoors times -0.15 (shown as winter use) plus the proportion of water used for sprinkling times -1.57. For this example, assume that the proportion of water used indoors is $\frac{8}{10}$, while the proportion of system water used for outdoors purposes is $\frac{2}{10}$. Then the overall price elasticity is -0.434, or $(-0.15 \times 0.8) + (-1.57 \times 0.2)$. This could be further adjusted for the price difference between the utility's \$2.00 price and the table price of \$1.68, yielding an elasticity of -0.517 or $(-0.434 \times \$2/\$1.68)$.

Under a third alternative, the estimated price elasticity for Kentucky from Table 9-5 of -0.92 could be used. In this case, by assuming the elasticity is proportional to price, multiply the coefficient of -0.92 by the ratio of the utility's \$2.00 price to the price in 1990 dollars from the Kentucky study, which is \$6.79, yielding $-0.271 = -0.92(\$2.00/\$6.79)$.

The three possible price elasticity values derived from these tables are -0.32, -0.52, and -0.27. These values are similar and average -0.37, which could then be used in the forecasting model. Values for other areas would not necessarily be this close together. Nevertheless, their average provides a reasonable approximation to the true elasticity.

Estimating Quantity Impacts and the Revenue Target

With a price elasticity of demand of -0.37, the 8 percent real price increase will decrease water demand by 1,184,000 gpd, or 40 mgd $\times$ 0.08 $\times$ -0.37. That is, the current usage is multiplied by the proportion of the price increase times the elasticity (Eq 9-2). The new sales will be 38,816,000 gpd, 2.96 percent below the unconstrained forecast.

The corresponding increase in revenue is the price elasticity of revenue multiplied by the proportion of the price increase times the current revenue (Eq 9-3). The price elasticity of revenue is 0.63, equal to 1 plus the price elasticity of demand (1 – 0.37). Thus, the increase in revenue is \$4,032, or 0.63 × 0.08 × \$80,000. The new total revenue is thus \$84,032/day.* This can be compared to the unconstrained forecast of \$86,400/day and revenues before the price increase of \$80,000/day (\$0.002/gal × 40 mgd). Thus, instead of increasing by the expected 8 percent or \$6,400/day (in real terms), revenues increased only 5.04 percent or \$4,032, an error of 37 percent!

Given the price elasticity of revenue, the price increase required to obtain any specified revenue increase (in real terms) can be directly computed. The required percentage increase in price is the desired percentage increase in revenue divided by the revenue elasticity coefficient (Eq 9-4). Thus, to obtain an 8 percent revenue increase in this example, simply divide 8 percent by 0.63, yielding a 12.7 percent increase in real price. Assuming next year's rate of inflation is 5 percent, a 17.7 percent nominal increase in rates and charges can be projected to yield an 8 percent real increase in next year's revenues in this system.

Cautions in Applying Elasticities

Two qualifications should be noted when using price elasticities. *Price increase* in this context means a proportional increase in all rates and charges. This interpretation is dictated by problems surrounding the relevant price specification and estimation of statistical price or rate impacts. Even when price increases are not strictly proportional, however, elasticity computations may still provide useful estimates.

Secondly, elasticities are only strictly valid for small changes in price. The larger the proportional price change, the wider the confidence interval must be around the estimated effects. Put another way, as the proportional price changes become larger, our confidence in the results necessarily decreases. For an extreme example using a price elasticity of demand of –0.5 and a 200 percent price increase (e.g., price goes from \$0.30/thou gal to \$0.90/thou gal), the simple computation outlined above would predict that water use would drop by 100 percent—to 0!

One simple solution to this problem is to do the computations incrementally. For that same 200 percent price increase, carry out the elasticity-based computations with a series of 10 percent increases. Thus, the first computation would increase prices by 10 percent with a resulting 5 percent drop in quantity. With compounding, the second 10 percent increase results in a total increase of 21 percent in price and a total drop in quantity of 9.75 percent. After 12 iterations, the total increase in price amounts to 214 percent while quantity falls by 46 percent. This procedure is approximately equivalent to assuming a double logarithmic functional form for the demand curve.

---

*Since elasticities are only exact for very small changes in price, the new total revenue computed using the revenue elasticity will differ somewhat from a computation using the new price and the new quantity. As the size of the price change approaches 0, this difference will disappear. As a practical matter, this difference will be much smaller than the random error in the forecast of revenues and can be ignored.

Summary

The diversity of rate schedules, including decreasing and increasing block rate schedules, suggests some problems regarding the definition or identification of the price of water. The following four main price specifications are noted in this discussion: the average price, the marginal price, the two-part price specification, and the total bill. Evidence from (1) controlled comparisons, (2) analysis of flat rate versus metered customers, and (3) econometric studies suggests that final resolution of the price specification problem is not necessary in order to demonstrate the existence of consumer responses to changes in water rates.

Because price elasticities are low and inflation typically erodes the impact of price increases, and because water is relatively cheap,* rate effects are underplayed and sometimes ignored in water resource planning discussions. When water is unmetered, or rates are very low, modest price changes can probably be safely ignored (Wong 1972; Houtakker and Taylor 1970). Nevertheless, rate or price effects are important since water rates and charges are usually the primary source of revenue for the water utility, and large rate increases usually cause a drop in sales.

The bottom line is that even relatively moderate price responsiveness in the urban water system can produce large errors in estimates of revenue when price effects are ignored. Thus, in the example computation previously discussed, increasing rates and charges by 8 percent produced additional revenues of only 5 percent (in real terms), an error of 37 percent!

This chapter suggests a relatively simple method for estimating price impacts of rate changes on urban water demand. While these computations will not provide perfect accuracy, water use forecasts that account for consumer response to price changes can be developed, and these forecasts will be more accurate than forecasts ignoring such effects.

Notes

1 The average price specification: Some analysts take the price of water to be the average price, or the consumer's total bill divided by the quantity of water purchased. When the fixed service fee is not too large, the average price reflects incentives associated with decreasing and increasing block rate schedules. Thus, the average price always declines for a decreasing block rate schedule. Average price increases for an increasing block rate schedule only when quantities are large enough that the effect of the increasing block rates more than offsets the monthly service charge.

The marginal price specification: Other analysts emphasize the marginal price, or water rate in the highest consumption block for a consumer in a billing period. For example, a consumer using 5,000 gal/month under the declining rate schedule of Table

*In 1990 residential water rates averaged less than \$2.00/thous gal, ranging from \$1.26 to \$2.34/thous gal. At these rate levels, water service to single-family residences cost between \$200 and \$400/year, or around \$20 to \$30/month.

9-1 faces a marginal price of \$2.50. The concept is that consumers optimize their situation and are concerned about whether the next increment of a certain good adds more or less enjoyment or use-value than its cost.

Marginal price and the "difference" variable: One school of thought depicts the impact of a rate schedule with two price or rate variables. One of these variables is the marginal rate. The second variable is the difference between the actual water bill and marginal price multiplied by total water purchased. For obvious reasons, this second variable in the two-part price specification often is called "difference." This specification is consistent with the theory of consumer choice in a full-information situation. When consumers know the exact rate schedule and also typical water consumption rates for household or commercial/industrial water-using activities, their response to changes in water rates is a combination of substitution and income effects captured by this two-part price specification.

The water bill: Finally, there is the view that the bill is the key consideration determining consumer responses. This is compelling in light of surveys that show that most water customers have no idea what the rate schedule looks like, but are at least somewhat aware of their monthly bill.

Calculating these "price" variables: Calculating the relevant prices or price proxies under these various price specifications shows several ways in which the changes in the rate schedule can give consumers incentives to consume greater or lesser quantities of water. With the declining block rate schedule in Figure 9-2, a consumer using 8,000 gal would pay a total bill of \$25.50 (\$5.00 service fee + 3 × \$3.00 + 3 × \$2.50 + 2 × \$2.00). This person's average price would be \$25.50/8 = \$3.19/thous gal. The marginal price (paid for the last unit of water purchased) is \$2.00. The two-part price variable, therefore, involves a marginal price of \$2.00. If the consumer purchased the entire 8,000 gal at this price, the outlay would be \$16.00. However, the actual bill is \$25.50. Hence, the second price variable in this specification—difference—is \$9.50. This is an "income effect" that makes the consumer feel poorer than if all water could be purchased at the marginal price.

Note that under the increasing rates of Figure 9-3, the consumer would pay \$25.80 (\$5.00 + 3 × \$2.20 + 3 × \$2.60 + 2 × \$3.20), for an average price of \$25.80/8 = \$3.225, and a marginal price of \$3.20. The "difference" variable in the two-part price specification would now be \$25.80 – (\$3.20 × 8) = \$0.20. For a customer using a larger volume, the difference variable would be negative. For lower volumes, the average price under decreasing block rates is higher than under increasing block rates. Conversely, for higher volumes, the average price is higher under increasing block rates.

2 Unless otherwise stated, a *price change* is understood in this chapter to mean a proportional change in all rates and fees. With this convention, the percentage change in any rate or water commodity charge, the total bill, and the average price are all the same for any given quantity of water. This *proportional change* interpretation is consistent with standard utility practices associated with rate increases, although in recent years, more emphasis has been placed on "rate restructuring."

3 R. Porges in *Factors Influencing Per Capita Water Consumption, Water and Sewage Works*, 1957, pp. 199–204, examines data from 1,474 municipal water systems,

determining an overall reduction in per capita water use from metering of approximately 25 percent.

4 A popular alternative functional form is multiplicative, which is often referred to as double logarithmic because it is typically estimated by taking natural logarithms of both sides of the equation. Thus,

$$Q = b_o P^{b_1} X_2{}^{b_2} \ldots X_n^{b_n}$$

becomes

$$ln(Q) = b_0 + b_1\, ln(P) + b_2\, ln(X_2) \ldots + b_n\, ln(X_n)$$

Multiplicative forms may fit a nonlinear relationship better than the linear form in Eq 9-1. There is the added bonus that the estimated coefficients—the a_1, a_2, etc.—are directly interpretable as elasticities that remain constant over the entire estimated demand curve.

A third functional form for the estimating equation was used to obtain the results shown in Table 9-4. It is

$$Q = b_0\, e^{b_1 P} X_2{}^{b_2} \ldots X_n^{b_n}$$

It can also be estimated by taking natural logarithms of both sides of the equation. The elasticity estimate from this model depends on price, and is $e = a_1 P$. Thus, the elasticity estimate is forced to increase in proportion to the price of water. This form is often referred to as inverse semi-logarithmic.

5 Note that the inclusion of a price variable P (or a suitable proxy) makes Eq 9-1 a demand function. The downward-sloping demand function, which readers may recall from economics in college, could be generated by Eq 9-1 by graphing it with price P on the vertical axis and quantity Q on the horizontal axis. All the other explanatory variables become "shifters" that move this downward sloping curve around the positive quadrant. The downward or negative slope of the demand curve comes from the typically negative sign of the coefficient of P, which is b_1 in Eq 9-1.

References

Babin, F.B., C.E. Willis, P.G. Allen, and A.S. Vlachou. 1981. Substitution Possibilities for Water Inputs in Selected Industries, NTIS PB81-209090.

Billings, R.B., and D.E. Agthe. 1980. Price Elasticities for Water: A Case of Increasing Block Rates. *Land Economics*, 56(2):73–84.

Billings, R.B., and W.M. Day. 1982. Price, Socioeconomic, and Climatic Influences on Residential Water Use in Southern Arizona, Project Completion Report. Tucson, Ariz.: Office of Water Resources and Technology.

Clark, R.M., and H.C. Goddard. 1977. Cost and Quality of Water Supply. *Journal AWWA*, 69(1):13–15.

Danielson, L.E. 1979. An Analysis of Residential Demand for Water Using Micro Time-Series Data. *Water Resources Research*, 15(4):763–767.

DeRooy, J. 1974. Price Responsiveness of the Industrial Demand for Water. *Water Resources Research*, 10(1):403–406.

Foster, H.S., and B.R. Beattie. 1979. Urban Residential Water Demand for Water in the United States. *Land Economics*, 55:43–58.

Gardner, B.D., and S.H. Schick. 1964. Factors Affecting Consumption of Urban Household Water in Northern Utah. Bulletin 449. Agricultural Experiment Station. Logan, Utah: Utah State University.

Gibbs, K.C. 1978. Price Variable in Residential Water Demand Models. *Water Resources Research*, 14(1):15–18.

Gottlieb, M. 1963. Urban Domestic Demand for Water: A Kansas Case Study. *Land Economics*, 39(2):204–210.

Griffin, R.C., and C. Chang. 1990. Pretest Analysis of Water Demand in Thirty Communities. *Water Resources Research*, 26(10):2251-2255.

Grima, A.P. 1973. The Impact of Policy Variables on Residential Water Demand and Related Investment Requirements. *Water Resources Bulletin*, 9(4):703–710.

Grunewald, O.C., C.T. Haan, D.U. Debertin, and D.I. Carey. 1976. Rural Residential Water Demand: An Econometric and Simulation Analysis. *Water Resources Bulletin*, 12(5):951–961.

Hansen, H.D., and R. Narayanan. 1981. A Monthly Time Series Model of Municipal Water Demand. *Water Resources Bulletin*, 17(4):578–585.

Houthakker, H.S., and L.D. Taylor. 1970. *Consumer Demand in America*. 2nd ed. Cambridge, Mass.: Harvard University Press.

Howe, C.W. 1982. The Impact of Price on Residential Water Demand: Some New Insights. *Water Resources Research*, 18(4):713–716.

Howe, C.W., and F.P. Linaweaver Jr. 1967. The Impact of Price on Residential Water Demand and Its Relation to System Design and Price Structure. *Water Resources Research*, 3(1):13–32.

Linaweaver, F.P. Jr., J.C. Beebe, and F.A. Skrivan. 1966. Data Report of the Residential Water Use Research Project. Baltimore, Md.: Johns Hopkins University, Department of Environmental Engineering.

Muncur, J.E.T. 1987. Urban Water Pricing and Drought Management. *Water Resources Research*, 23(3):393–398.

Rensetti, S. 1988. An Econometric Study of Industrial Water Demands in British Columbia, Canada. *Water Resources Research*, 24(10):1569–1573.

Weber, J.A. 1989. Forecasting Demand and Measuring Price Elasticity. *Journal AWWA*, 81(5):57–65.

Wong, S.T. 1972. A Model of Municipal Water Demand: A Case Study of Northeastern Illinois. *Land Economics*, 48(1):34–44.

Chapter 10

Water Conservation

Utilities frequently implement several water efficiency programs and measures simultaneously, thus synergism and competition between programs can become an issue in projecting system water savings. Often, the best approach is to conduct on-site statistical research, either through control group comparisons or with regression models, using a scientific sample of data on customer water use. Detailed discussions of these approaches to estimating water savings potential are presented in this chapter. The results of a nationwide survey of the use of water conservation efforts provide an introduction to these programs.

Types of Conservation Programs

Based on a questionnaire distributed to a sample of 1,000 public water systems in American Water Works Association (AWWA) files, an estimated 175 million people were served by utilities implementing some form of water efficiency or conservation program during 1990.*

Major types of programs involve

- public information
- free or low-cost distribution of toilet or shower flow restrictors

---

*The water demand questionnaire obtained responses from public water supply systems serving some 48 million people. Because these respondents were part of a stratified random sample, it was possible to extrapolate from the sampled patterns to the population coverage estimate of 175 million people. See Jones (1992).

- higher water prices for peak demand periods
- plumbing codes specifying low-water-use fixtures or appliances
- use of treated effluent for outdoor purposes
- short-term restrictions on water use at certain times of the day or week (e.g., for lawn watering and car washing)
- efficient irrigation systems
- promotion of low-water-use landscaping (Xeriscape)
- landscape ordinances requiring low-water-use plantings
- monitoring customer compliance during droughts

Table 10-1 summarizes the relative popularity of these programs as of 1990. California public water utilities show dramatically higher participation rates for these water conservation/efficiency programs than elsewhere in the nation.* While most California utilities distributed free or low-cost toilet and shower flow restrictors, only 68 percent of all surveyed utilities nationwide distributed toilet-tank displacement devices and only 29 percent offered shower flow restrictors.

The water demand questionnaire also polled utility managers on reasons for implementing water efficiency programs. Five major reasons were water shortage (32 percent), systems operations or capital cost savings (17 percent), regulatory requirements (14 percent), public interest group pressure (4 percent), and declining groundwater table (3 percent).

Utility managers credit water conservation/efficiency programs with reducing water use in the short and long term. Respondents supplied water use savings estimates by specific measure and, in a few cases, for the utility's combined conservation program. Information programs, including advertising and public service announcements on radio or television and in the newspapers, produced short-term average system water savings of 1–2 percent. Water conservation devices, including low-flow showerheads, ultra-low-flow toilets, toilet dams, and other low-flow fixtures, resulted in estimated average short-term savings of 2–9 percent. Measures such as across-the-board rate increases, changing from a declining block rate schedule to a constant commodity charge, going to an increasing block rate schedule, and imposing penalties for overuse accounted for short-term average savings of between 1 and 20 percent, short-term peak reductions of from 2–25 percent, and long-term average reductions of between 2 and 20 percent. Finally, water restrictions were estimated to reduce average water use by 5–30 percent and peak water use by 6–50 percent in the short term, and to account for 2–30 percent reductions as a long-term average. A handful of utilities supplied estimates of overall water savings from their combined programs of between 12 and 25 percent.

---

* Since larger public water supply systems tend to implement more water efficiency programs, the percentages in the third column of Table 10-1 may exaggerate differences between California and non-California patterns. Nevertheless, water efficiency initiatives in California dating back into the 1980s and the prolonged drought appear to have increased public water supply program participation in that state.

Table 10-1 Implementation of water efficiency programs in the United States

	Percent of Sample PWS* Implementing Program	Population in National Sample Served by PWS Implementing Program, in *millions*	Percent of California Sample Population Served by PWS Implementing Program
Information programs	68	37.4	100
Free or low-cost distribution of toilet or shower flow restrictors	29	19	98
Higher water prices for peak demand periods	13	10.4	43
Building codes specifying low-water-use appliances or landscaping	25	18.1	64
Rebates for low-flow toilets or water-conserving landscaping	11	10.9	70
Use of treated effluent for outdoor purposes	10	9.5	75
Restrictions on water use during periods of the day/week	26	16.2	63
Customer monitoring and enforcement	15	10.7	81

NOTE: Percents and populations for PWS implementing various programs are counts from the water demand questionnaire.

* *Public water suppliers.*

Evaluating Water Conservation Programs

Most analysts suggest that water conservation or efficiency programs, policies, practices, or measures should be evaluated using a "net benefits" criterion to determine if the program is worth its cost.* A comprehensive evaluation of urban water conservation programs (Planning and Management Consultants et al. 1991) lists seven standard methods

*An early, landmark study of water conservation states, "... the first axiom of the definition of water conservation is that the beneficial effects of the reduction in water use (loss) must be considered greater than the adverse effects associated with the commitment of other resources to the conservation effort." (See Bauman et al. 1978.)

of economic analysis that can be performed to evaluate relative costs and benefits of urban water conservation programs. The methods include

- net present value
- benefit-to-cost ratio
- internal rate of return
- discounted payback
- life-cycle revenue impacts
- levelized costs and benefits
- number of years to recover capital outlay

The universe of water conservation and efficiency programs that may or may not meet the net benefit criterion under one or more of these methods in specific water districts is extensive. A useful list of conservation measures was compiled in Maddaus (1987) and is shown in the box on the opposite page.

To facilitate the process of choosing an actual program, measures with expected positive net benefits can be ranked according to social acceptability based on local values, practices, and experience. There is typically broad consensus about the value of (1) pursuing leak control in the water system, (2) providing incentives to customers to install and use low-water-use appliances and fixtures, and (3) installing in-house recycling equipment for industrial cooling systems. Pricing policies are often politically controversial, but can be effective. Many of the pricing policies are designed to reduce the peak level of demand, since peak demands often contribute to escalating system capital and operating costs. Direct controls in the form of water restrictions or ordinances limiting new construction usually are short-term solutions, due to public dislike of measures designed to compel behavior.

Incorporating Conservation Into the Forecast

The basic method of including water conservation and efficiency savings in demand forecasts is to deduct the estimated savings associated with each measure or program from system water demand. In a regression-estimated structural/causal model, however, the conservation measures are represented by independent variables and their effects directly incorporated into the results. Savings estimates can be obtained from engineering end-use studies or local statistical analysis of water conservation and efficiency programs. In some instances, total savings may be less or greater than the sum of the individual water savings associated with each conservation measure, practice, or program implemented by the water utility. One task in integrating conservation savings estimates into forecasts, therefore, is to make reasonable allowance for overlapping and complementary aspects of water efficiency programs.

Synergism, Competition, and Independence

Water Conservation Measures

General: public information, in-school education, metering, pressure reduction, pricing policies (uniform commodity rates, increasing block rates, seasonal rates), leak detection and repair, system rehabilitation

Interior residential use: low-flow showerheads, shower flow restrictors, toilet-tank inserts, pipe insulation, faucet aerators, efficient appliances, low-flush and ultra-low-flush toilets

Power generation: recirculation of cooling water, reuse of treated wastewater, in-system treatment

Industrial use: recirculation of cooling water, reuse of cooling and process water, reuse of treated wastewater, efficient landscape irrigation, low-water-using fixtures, process modification

Landscape irrigation/design: efficient landscape design, low-water-use plant material, scheduled irrigation, efficient irrigation systems, tensiometers, reduction or limitation of high-water-use plant materials, such as turf

Source: Maddaus (1987).

Often, several water conservation or efficiency programs are implemented at the same time, e.g., a publicity campaign, a rebate program, and a rate increase. When this is the case, it is necessary to consider *synergism* and *competition* (antagonism) between the measures implemented in order to derive reliable estimates of overall water savings.

A *synergism* is when the total water savings from implementing several water efficiency or conservation measures is greater than the sum of the expected water savings associated with each measure. An example is the interaction of pricing with other conservation measures. Higher water prices encourage consumers and producers to take advantage of the full potential for water savings from available measures or devices (see Berk et al. 1981). Thus, using price and nonprice measures independently is expected to provide a conservative estimate for programs including price and nonprice measures, provided that the price elasticity as well as the nonprice effects are measured in an appropriately specified model. Insufficient research has been carried out, however, to ascertain the size of these effects with any certainty.

Competition or antagonism between water conservation measures, practices, programs, or policies exists when total water savings are less than would be indicated by adding up the expected individual water savings. An example might be an ordinance requiring all new showerheads sold to be low flow and a showerhead retrofit program distributing free flow restrictors. The sum of the water savings from each program, taken by itself, would be greater than their combined effect, since some of the same homes would be reached by both programs.

A thoughtful examination of the source of the competitive effect on the total savings estimates will often suggest an immediate solution. In the preceding example, the number of homes in the retrofit program could be reduced by an estimate of how many homeowners are likely to replace their showerheads each year.*

*This basic point is made with a geometric argument in Stein and Kilgore (1990).

Estimating Water-Savings Potential

Case studies, success stories, and anecdotes help establish a context for and expectations about water savings from water conservation programs. Fresno, Calif., for example, reportedly reduced its system water demand by 13 percent during a peak demand period in 1974 using a public education program.* Free distribution of water-saving devices was credited with saving a little more than 3 percent of average household water use in Oxnard, Calif., in mid-1977 (Morgan and Pelosi 1980).†

Table 10-2 presents estimated water savings for a variety of conservation measures. Many of the water-savings estimates in Table 10-2 rely on the technical characteristics of the devices or programs and anticipated customer participation rates.

These estimates support what is often called an *engineering*, or *end-use, approach* to estimating water savings from a conservation program. Total water savings are represented in this approach as a product of several factors, including (1) the water saving per use of a device, appliance, or fixture, (2) the frequency of use of the device, appliance, or fixture per time period, (3) the potential market for acceptance of measures in terms of the number of water customers, and (4) the fraction of water customers who are likely to accept the measure and install the device in each time period to be analyzed.

Thus, end-use estimates, although founded on technical characteristics of water-using devices, require information about how people behave in connection with this equipment. Also required is a projection of how the equipment will be accepted in the service area of the water utility. Finally, the estimated savings depend on customers refraining from offsetting changes in behavior, such as using the water saved by a low-flush toilet to irrigate additional flowers or trees.

Variability of Conservation Program Impact

The scope and content of water conservation and efficiency programs varies widely by location and by time period in the same location. Initial public acceptance of measures such as shorter showers and reduced lawn sprinkling may erode over time as people revert to "normal" water use patterns. There is some evidence that people may respond to the use of shower flow restrictors or low-flow showerheads by taking longer showers (Geller, Erickson, and Buttram 1983; Hamilton 1983). Temporary flow restrictors in showers and faucets, and water displacement devices in toilet tanks may be removed when these fixtures are serviced a few years after their initial installation.

Water Conservation Goals

When water conservation results are estimated before the programs are implemented, actual results may vary significantly from what was expected. A specific long-term conservation goal can be achieved, however, when utility officials commit themselves to that goal and retain the flexibility to implement a variety of programs as

---

* Personal communication, Peter Macy.

† Two good sources for estimates on water savings include *Water Conservation*, distributed by AWWA; and *Residential Water Conservation Projects—Summary Report*, produced for the US Department of Housing and Urban Development by Brown and Caldwell in June 1984.

needed to achieve it. Extensive on-site studies of the effectiveness of water conservation programs indicate water savings within the 1–20 percent range are feasible for the long term (Texas Water Development Board 1991; California Department of Water Resources 1991). Whether such reductions are desirable is a matter for local determination, based on the relative costs and benefits involved. Given, for example, a utility commitment to achieve consistent 10 percent average savings over the period of a decade, such savings should be achievable with the range of measures, practices, programs, and policies that have been tried and proven in the urban water industry to date. The process involved is illustrated in Figure 10-1.

Table 10-2 Engineering end-use estimates of water savings

Water Efficiency/ Conservation Measure	Water Savings
Bathroom	
1.0-gal-per-flush (gpf) toilet	18.0 gal per capita per day (gcd)
1.6-gpf toilet	15.6 gcd
Toilet flush valve retrofit	2.0 gpf
Urinal flush valve retrofit	2.0 gpd
1.5-gal-per-minute (gpm) showerhead	9.9 gpd
2.75-gpm showerhead	7.2 gcd
Faucet aerator	0.5 gcd
Toilet-tank displacement bag	2.8 gcd
Toilet-tank displacement dam	4.0 gcd
Toilet leak repair	4.1 gcd
Outside Irrigation	
Drip irrigation	20% of exterior use
Soil preparation/mulching	15% of exterior use
Landscape requirements	20% of exterior use
Low-water-use landscape	50% of exterior use
Moisture sensors	40% of exterior use
Operational	
Leak detection and repair	10% of unaccounted-for water
Universal metering	20% of system unmetered usage
Wastewater Ordinances	
Evaporation reduction (e.g., irrigation in morning or evening)	30% of exterior use
Grey water use	15% of exterior use

Sources: Residential Water Conservation Projects—Summary Report *(available from the US Department of Housing and Urban Development); Maddaus (1987).*

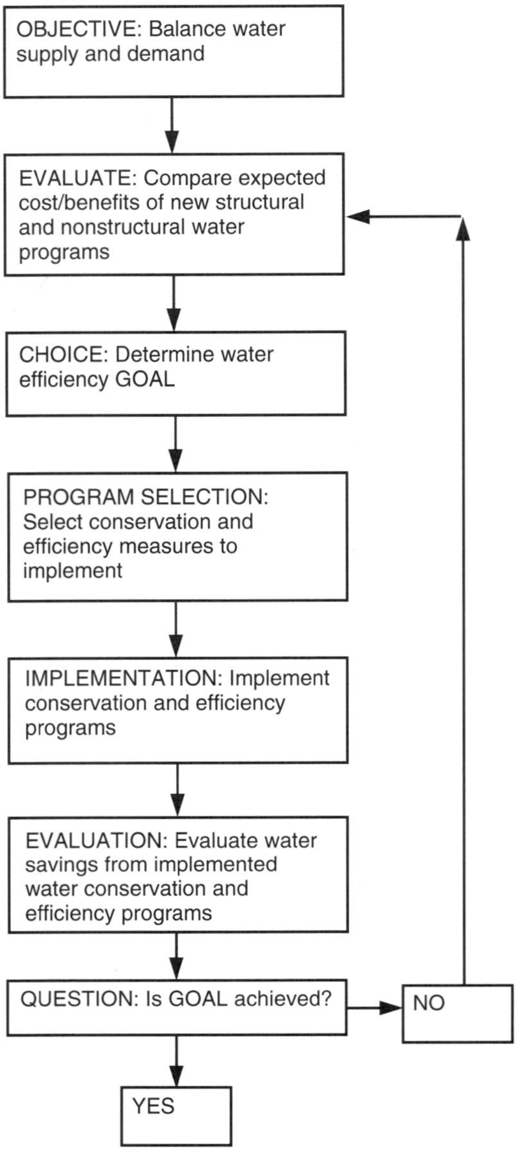

Figure 10-1 Flow diagram of goal achievement process

Estimating Conservation Effects With Local Studies

The wide variability in the reported impacts of conservation programs suggests the need for local statistical study of water conservation and efficiency programs, and continued monitoring of their effectiveness. The simplest approach uses a comparison of aggregate weather-normalized water use before and after implementation of a conservation program. For more detailed information, the response of a sample of individual customers to a conservation program can be evaluated using control group comparisons or a regression-estimated demand model.

In order to ensure accurate monitoring of a conservation plan, all three methods mentioned require careful planning prior to the implementation of the conservation program.

Aggregate Before and After Analysis

When one or more conservation programs are implemented during a relatively short period of time, their impact can be evaluated using an aggregate water demand model. Water use before implementation of the program can be normalized for weather and projected into the future to be compared with actual water use after the program has been fully implemented. This works effectively for relatively short time intervals, since the analyst can ignore changes in other longer-term factors such as income, housing patterns, commercial activity, etc. If some nonconservation factor is known to have changed, it should be built into the model. Methods of weather normalization are presented in chapter 6.

The normalized water use data from before and after the implementation of a conservation program can be used to determine the size of any resulting change in water use and the chance that this change did not result from the program but from random chance. The significance of a measured change in water use can be computed using a t statistic.

Suppose that water use is measured and normalized for seasonality, weather, and trend (if any) during 12 months before and another 12 months after the implementation of a new conservation program. Assume that normalized water use averages 4,500 mil gal per day (mgd) before and 3,900 mgd afterwards. From these data, the standard deviation before (S_1) is computed as 450 mgd and afterwards (S_2) at 520 mgd. With 11 ($n-1$) degrees of freedom, the critical t statistic at the 5 percent level for a one-tailed test is 1.796. The test statistic for the change in water use is

$$t\sqrt{\frac{S_1^2}{n_1}+\frac{S_2^2}{n_2}} = 1.796\sqrt{\frac{450^2}{12}+\frac{520^2}{12}} = 356.5 \qquad (10\text{-}1)$$

Since mean normalized water use changed by 600 mgd, the change in water use is statistically significant at the 5 percent level. This means there is less than a 1 in 20 chance that the observed drop in water use was due to random change rather than the program. In this example, there is actually less than a 1 in 100 chance that the drop in water use was due to random chance, since the critical t statistic at 1 percent is 2.718.

This aggregate before-and-after method of evaluating a conservation program is the simplest and least expensive of the three approaches discussed. It is especially valuable for assessing the overall impact of a conservation program. Its primary limitation is that no information on individual responses to the program is provided. The detailed data on individual customers, which would be collected for the other approaches, may be valuable in designing program changes.

Control Group Comparisons

Control group comparisons involve a "with and without" analysis. One group of water customers is selected to use a water conservation device, or otherwise become actively included in a specific water efficiency program. This is called the *treatment group*. A second group, called the *control group*, is excluded from these initiatives.

Differences in the water use of the treatment and control groups are measured before and after participation in the program. Using random sampling to select the comparison groups creates a basis for standard tests to determine whether (1) observed differences in the water consumption of the treatment and control groups are due to the program under study or (2) whether such differences could be explained as chance variations.

Example Analysis

Suppose, for example, that a 2,000-customer treatment group averaged 60,000 gal/household before treatment and 55,750 during the study period. Meanwhile, the 2,000-customer control group average was 56,000 gal/household before and 52,200 during the experiment. Thus, water use fell by an average of 4,250 gal in the treatment group, but by only 3,800 gal in the control group. Is the 450-gal difference in these changes statistically significant or can it be explained as a chance variation?

The statistical test is based on the t statistic. The test accepts or rejects the null hypothesis, which is that the observed difference in the group averages is due to chance. A one-tailed test is appropriate in this context, since we are interested in whether or not the difference represents a true water savings, i.e., water savings due to the treatment is significantly greater than 0.* The critical t value for the one-tailed test taking a 5 percent chance of rejecting the null hypotheses when it is true is 1.645. The test relies on the standard deviations estimated from the sample data. For this example, assume the standard deviations of water use reduction for the treatment and control groups, each of which has 2,000 members, are respectively 2,800 and 2,100 gal. The test statistic is

$$t\sqrt{\frac{S_1^2}{n_1}+\frac{S_2^2}{n_2}} = 1.645\sqrt{\frac{2{,}800^2}{2{,}000}+\frac{2{,}100^2}{2{,}000}} = 128.74 \qquad (10\text{-}2)$$

Since the differences in the water use reduction between the treatment and control group is 450 gal, which is greater than the critical test statistic of 128.74 gal, we can reject the null hypothesis of no difference and accept the alternative hypothesis that the treatment did result in a statistically significant reduction in water use.

Setting Up a Controlled Comparison

Suppose a water utility with 9,000 residential customers is considering plans to distribute and install shower flow restrictors at no cost to customers and wants to run a pilot study to determine the likely water savings. Suppose also, as in the summer residential water use distribution presented in chapter 2, the population standard deviation of summer residential water use is approximately 5,000 gal for the July–August billing period and the average customer water use is 8,000 gal. End-use estimates suggest that shower flow restrictors can save 300 gal in a typical household over a two-month period. Assuming all treatment group households elect to install the showerheads, and that none of the sample households are already using these devices, what sample size is necessary to estimate, at a 95 percent confidence level, the actual decrease in average water use?

---

*We are not interested in the opposite "tail" of the test, which would be that the conservation program increased water use.

The key equation involves the tolerance, which is 300 gal; the standard deviation of water use, which is 5,000; and a t statistic of 1.645, which was obtained from a table for the normal distribution for a 95 percent one-tailed confidence interval. The required sample size is

$$N = \left(\frac{tS}{tolerance}\right)^2 = \left(\frac{(1.645)(5{,}000)}{300}\right)^2 = 752 \qquad (10\text{-}3)$$

This can be slightly reduced with the "finite sample correction" (FSC), which is

$$FSC = \sqrt{\frac{population - sample\ size}{population - 1}} = \sqrt{\frac{9{,}000 - 752}{9{,}000 - 1}} = 0.957 \qquad (10\text{-}4)$$

Accordingly, it would be necessary to include 720 customers (0.957 × 752) in the treatment group receiving the device and another 720 in the control group, to obtain a controlled comparison of average water use with and without the device.*

Once the sample size is determined, a random sample of customers is selected and randomly assigned to the treatment and control groups. The conservation devices would then be installed in the houses of those assigned to the treatment group. Households that already have water-saving showerheads or flow restrictors would be excluded from the sample and replaced with additional randomly selected households.

Ideally, placebo devices would be installed in the homes of the individuals assigned to the control group. The members of the groups would not be told which device they received. This procedure may seem cumbersome, but it avoids the all-too-real problem of contaminating the results because the individuals receiving the devices want to please the research team by obtaining the anticipated results.

An obvious problem is that some households will refuse to participate in the experiment. At best, this requires starting with a larger sample to achieve the target sample size. At worst, it partially invalidates the results, since nonparticipants may be different in their water use behavior than participants. Refusals do, however, provide valuable information on the likely penetration rate of a utility-wide program. There are many provisos to setting up and conducting a reliable controlled comparison (see Planning and Management Consultants 1991).

The water use of all customers in both the treatment and control samples is measured before, during, and after the experimental period. Then, it is possible to compute the sample standard deviations for the treatment and control groups, and t-tests for the difference of means can be applied. This experiment would provide an estimate of the mean and standard deviation of water saved by the device, as well as an estimate of the penetration rate that could be expected for the entire utility given that every customer was requested to install the device with the same intensity as the sample.

---

*Further adjustments can be made to this figure based on the anticipated participation rates of customers selected for the treatment group.

Disaggregated Causal/Structural Model

Water demand models relating water consumption to determinants of water use, such as household income, water price, age of occupants, household size, and new landscaping, or other factors, also can be used to estimate the effects of water conservation programs. A multivariate, cross-sectional modeling approach is often used to estimate water use for individual accounts. The same demand model would incorporate explanatory variables representing conservation programs, so that their effect on water use could be estimated.

For example, in a study of monthly summer water use at apartments with grass in Tucson, Ariz., the following linear demand model was estimated by multiple regression:

$$\begin{aligned} Q = 6.22 + 0.85\,YARD + 1.00\,SHOWER \\ -2.09\,FAUCET + 0.70\,KNOWLEDGE - 1.75\,WC \\ + 0.12\,VACANT - 0.014\,SIZE + 0.71\,POLICY \end{aligned} \tag{10-5}$$

Where:

Q = average summer water use per occupied apartment per month in hundred cubic feet (Ccf). Mean water use was 10.5 Ccf/apartment.

YARD = type of yard, ranging from 1 = desert to 5 = lush vegetation

SHOWER = use of water-saving showerhead = 1, regular showerhead = 0

FAUCET = low-flow faucets = 1, otherwise = 0

KNOWLEDGE = rate structure knowledge, ranging from 0 = none to 5 = maximum

WC = water-saving toilet or toilet-tank dam = 1, otherwise = 0

VACANT = vacancy rate, in percent

SIZE = size of complex. Size varied from 2 to 456 units

POLICY = managers have authority to implement water-conserving changes = 1, no authority = 0

Water use is expected to increase with increasing values of *YARD*. The coefficients on each of the three water-saving fixtures are expected to be negative, indicating that their use does reduce water consumption. Water use is expected to decrease with manager knowledge of water rates, so the coefficient on *KNOWLEDGE* should be negative. Likewise, the expected sign on *POLICY* is negative, since apartments where managers have authority to institute water-saving changes can be expected to use less water. The coefficient on *VACANT* is expected to be positive, since water use is defined per occupied apartment. Thus, irrigation water use is spread over fewer apartments when there are numerous vacancies. Landscaping uses the same amount of water regardless of occupancy. If water use were defined as water use per apartment (vacant or occupied), the sign on *VACANT* would be negative. Finally, the sign on *SIZE*, the number of apartments in a complex, might be expected to be negative, reflecting economies of scale in water use in apartment complexes—perhaps because of the decrease in the irrigated yard area per apartment in larger complexes.

The model was estimated using data from 178 apartment manager interviews done in January 1989 and water use data from the water utility for the summer of 1988. Since they all paid the same price for water, no price variable is included in this model. Additionally, since the apartments are all located in Tucson, they all experienced the same weather conditions, allowing for the omission of weather and climate variables. The data for this regression are provided on the data diskette, so the interested reader can replicate this result as well as explore the effect on the model of changing the model specification by omitting individual variables. The reader can also explore the effect of alternative functional forms for the model, such as double logarithmic.

The two most interesting results from the model are

1. The prediction that apartments with low-flush toilets will use 1.75 Ccf of water less per month in the summer than those without.
2. The prediction that low-flow faucets save 2.09 Ccf of water per month.

Before using results such as these, however, it is necessary to evaluate the reliability of the model and its results. Equation 10-5 has an R^2 of 0.136, indicating that 14 percent of the variation of water use from its mean value is explained by the model. Although this is relatively low, even for a cross-section model, the regression may still provide some useful results. The F statistic is 3.327, which is significant at 0.0015, meaning that the equation is highly significant from a statistical viewpoint. The standard error of the estimate is 6.046 Ccf. Unfortunately, the estimated savings from the water-saving devices is smaller than the standard error of the estimate, which means that the estimated savings are less than the typical error in the estimates.

Before using the results of this model, the analyst should attempt to reduce the size of this standard error. The large standard error may be due to errors in reporting by apartment managers or to other causes. If the analyst could physically measure the flow rates of installed showerheads and faucets, and measure the gallons required/flush for toilets, these potential errors could be eliminated. Resources seldom permit this type of detailed data gathering, however.

All of the individual estimated coefficients, with the exception of *SHOWER* and *POLICY*, were significant at 5 percent or better. Thus, there is less than a 1 in 20 chance that the true value of the coefficient is 0 or has the "wrong" sign (using a one-tailed test, since there is a stated hypothesis on the expected sign of each variable).

The estimated coefficients on *SHOWER* and *POLICY* have unexpected positive signs and are not statistically significant. These results may be due to errors in reporting by apartment managers or to other unspecified causes. Additionally, the estimated coefficient on *KNOW*, manager knowledge of water rates, has an unexpected positive sign. This variable is statistically significant at a 5 percent level using a two-tailed test, which would be used if the initial hypothesis about the direction of its effect were undetermined.

Analysts often drop insignificant variables from their models, and the reader can verify that dropping the two insignificant variables from this specification lowers the R^2 only slightly and slightly increases the adjusted R^2. Interestingly, when *SHOWER* and *POLICY* are dropped from the regression, the F statistic rises to 4.18. The relatively minor changes in the other coefficients when these two variables are omitted from the model increases our confidence in these results. The rule of parsimony, i.e., using the simpler model, supports dropping *SHOWER* and *POLICY* before using a model such as this for forecasting applications.

This regression illustrates the type of information that is available from collection and analysis of firsthand sample data on water use and conservation. The regression model validates the use of certain conservation savings estimates, such as those for low-flow toilets and water-saving faucets. In other respects, the regression analysis suggests certain problems and issues for further investigation. For example, why would more knowledge about rates lead to increases in water use? Do the positive coefficients on *SHOWER* and *POLICY*, together with the low statistical significance of these estimates, mean that low-flow showers and managerial authority to implement water conservation are just not important in the overall water use picture?

Clearly, validation of some end-use estimates and raising questions about other effects are potentially more valuable than simply using a list of end-use estimates for water conservation developed from national data or other water systems and not attempting to understand local conditions or attitudes about water use.

Summary

Water conservation or efficiency programs are increasingly important in urban water systems. Millions are served by systems with multiple water conservation efforts, including publicity about why conservation is important. The potential for water conservation is enormous, since only a small percent of delivered water is actually used directly by people. In Los Angeles, for example, only 11 percent of urban water is used by households for personal purposes (Johnson 1988).

A simple way to include water conservation and efficiency program effects in demand forecasts is to deduct the estimated savings of the conservation program from the unconstrained (without conservation) system water demand forecast. A more sophisticated approach would include conservation impacts as independent variables along with the price of water, personal income, business activity, weather, etc., in a multivariable forecasting model.

The broad range of estimates of conservation savings signals the difficulty of estimating the effectiveness of particular water conservation/efficiency programs before they are implemented. This difficulty is compounded by synergistic and competitive interactions among individual conservation programs. These uncertainties suggest the need for local statistical studies of water conservation programs, as well as continuing monitoring of their effectiveness.

The generally accepted approaches for studying the effectiveness of water conservation programs include (1) control group comparisons and (2) multivariate models estimated by regression analysis. Specific water use reductions in a range from 5 to 20 percent can be achieved by utility officials determined to use any or all available conservation program elements. In this case, the water conservation plan may become a highly accurate forecast (refer to Texas Water Development Board 1991; California Department of Water Resources 1991).

References

Baumann, D.D., J.J. Boland, J.H. Sims, B. Kranzer, and P.H. Carver. 1978. *The Role of Conservation in Water Supply Planning*, Contract Report 78-2. Institute for Water Resources, US Army Corps of Engineers.

Berk, R.A., C.J. LaCivita, K. Sredl, and T.F. Cooley. 1981. *Water Shortage: Lessons in Conservation from the Great California Drought, 1976–1977*. Cambridge, Mass.: Abt Books.

California Department of Water Resources. 1991. *Urban Drought Guidebook — New Updated Edition*. California Department of Water Resources, Water Conservation Office.

Geller, E.S., J.B. Erickson, and B.A. Buttram. 1983. Attempts to Promote Residential Water Conservation with Educational, Behavioral, and Engineering Strategies. *Population and Environment*, 6(2):96–112.

Hamilton, L.C. 1983. Saving Water—A Causal Model of Household Conservation. *Sociological Prospectives*, 26:355–374.

Johnson, D. 1988. Down the Drain. *Los Angeles Times*, V1, Feb. 2.

Jones, C.V. 1992. *Benchmarking Urban Water Efficiency Practices in the United States, 1990 Data, Appendix to Letter Report for Standard Methodology and Initial Studies for Estimating Reductions and Impacts of Reductions of Water Use in the U.S.* Submitted to the US Environmental Protection Agency by James M. Montgomery Engineers, September 1992.

Maddaus, W. 1987. *Water Conservation*. Denver, Colo.: American Water Works Association.

Morgan, W.D., and P. Pelosi. 1980. *The Effects of Water Conservation Kits on Water Use in Water Conservation Strategies*. Denver, Colo.: AWWA.

Planning and Management Consultants in cooperation with Brown and Caldwell Consultants, Spectrum Economics, and James M. Montgomery Consulting Engineers. 1991. *Evaluation of Urban Water Conservation Programs: A Procedures Manual, Part II*. Prepared for the California Urban Water Agencies.

Stein, S.M., and R. Kilgore. 1990. Interaction Between Water Conservation Measures and the Effects on Conservation Programs. *Proceedings of Conserve 90*. Denver, Colo.: AWWA.

Texas Water Development Board. 1991. *Guidelines for Municipal Water Conservation and Emergency Water Demand Management*. Texas Water Development Board.

Chapter 11

Forecast Uses, Evaluation, and Improvement

This chapter reviews the steps in water demand forecasting; provides descriptions and applications of short-, medium-, and long-term forecasts; and examines forecast interval estimates, forecast evaluation, error measurements and loss functions, strategies to improve forecasting accuracy, and the reasons that formal forecasting is valuable.

Forecasting Process

Forecasting starts with the realization that new information offers the potential of assisting utility officials to make better decisions. Steps in the forecasting process include

1. Clarification of the purposes of the forecast.
2. Choice of the forecast approach.
3. Collection and analysis of data.
4. Identification of forecast model(s).
5. Estimation of the forecast model(s).
6. Diagnosis of the statistical adequacy of the model(s).
7. Production of the forecast, including confidence intervals.
8. Evaluation of the forecast.
9. Use of the forecast by decision makers.
10. Ex post analysis of forecast error.

Short-Term Forecasts

Forecasts are often described by their relative length as short term, medium term, or long term. Short-term water demand forecasts have practical application in operating a utility. Decisions on maintenance and operating schedules for pumps, wells, reservoirs, and mains are partially based on short-term forecasts of demand. Reservoir operation decisions about water levels and degree of drawdown are also based on demand forecasts along with forecasts of available supply. All utilities use forecasts for these purposes, even if they are as simple and intuitive as, "We do our pump maintenance work in January, when demand is low."

Fluctuations in water use in the short run tend to be dominated by the weather. While the forecaster can adjust for the normal seasonal variation in water use from winter to spring to summer and fall, there is currently no reliable way to forecast water use variations from the "normal" weather pattern.

Simple trending and more complex, univariate time-series methods are the primary short-term forecasting techniques. The ease of use, low cost, and relative accuracy of simple trending models often make trending the first choice for short-term forecasting.* In many instances, the importance of budgetary decisions tied to short-term forecasts of water sales and revenues justifies the application of more complex, but more accurate approaches, such as exponential smoothing. As noted previously, exponential smoothing was one of the most successful forecasting methods in the international forecasting competitions held in the late 1970s and early 1980s (the "M-competition," see page 94).

The major limitation of extrapolation models is their reliance on historic patterns. The assumption that the future will be the same as the past works well as long as underlying conditions are changing in a uniform manner. However, when such key factors as price or conservation programs vary, forecasts from these models must be adjusted externally. This adjustment is based on judgment and knowledge of the probable impact of these changed conditions.

For some situations, a short-term forecast might be produced or supplemented with sample survey data from utility customers. Such a survey might ask respondents to describe their water-using activities, along with both recent changes in those activities and plans for the future. In this way, if numerous customers are planning to plant grass (or replace grass with gravel), the water utility can anticipate a corresponding change in water use.

These surveys might be most useful for industrial customers, who may have plans for expansion or modification of their production activities that would impact their future water use. A utility with a few large industrial water users may benefit from regular periodic surveys of these customers. Information should also be collected on the probable impact of changes in government regulations regarding water use or wastewater disposal.

Causal/structural models based on surveys of individual customers are useful in obtaining unbiased estimates of the impact of price, income, and other key explanatory variables in water demand. Often, such models have relatively low R^2 but strong t statistics

* Here, trending can be as simple as applying an expected growth rate to water sales.

(low standard errors) for individual regression coefficients. Such regressions can often provide relatively accurate estimates of the impact of certain explanatory variables as well as forecasts of overall sectoral water use (see chapter 5).

Long-Term Forecasts

Water utilities face a formidable long-run forecasting problem. The challenge is to have sufficient capacity ready to meet ever-changing customer demands while avoiding costly excess capacity. Errors in demographic forecasts, on which water demand forecasts are necessarily based, directly translate into insufficient or excess capacity, either of which imposes substantial political and economic costs on the utility and its managers. Sensitivity or risk analysis is often helpful in evaluating the potential range of required supply capability.

Long-term forecasts are sometimes completed using intuitive methods, such as Delphi (see appendix A, page 156). Qualitative issues may be particularly important in the longer run, especially those pertaining to directions in the local or regional economy. For many areas, the key questions are whether industrial decline will be reversed or vigorous economic growth reinstituted.

Per capita and unit use water demand forecasts are the primary long-range water demand forecasting techniques. Customer water use coefficient forecasts often can help identify the mix of trends applying to components of water use that tend to be masked in the aggregate.

The initial projection of an unconstrained demand forecast based on no changes in the economic/social environment can be modified to produce a constrained forecast with the use of environmental changes and appropriate elasticity estimates. When capacity constraints exist or expansion financed by higher water rates is necessary, unit use projections can be modified in an iterative fashion to develop a logically consistent forecast.

Medium-Term Forecasts

Medium-term forecasts cover a time span of from two to five years and are used to facilitate utility operations, minor system improvements, revenue forecasts, and rate setting. While many of the factors discussed with respect to long-term and short-term forecasting also apply to medium-term forecasts, the primary additional complication arises from cyclical behavior of the weather and the economy. Both are important influences on water demand, and both are subject to recurring multiyear fluctuations of uncertain and somewhat random severity and duration.

Predicting cycles in the weather is perplexing. It is well known that both dry and wet years tend to run in multiyear cycles. Additionally, the longer a drought (or wet period) has continued, the more likely it is to continue for another year. However, these multiyear cycles can and do change abruptly. About all the forecaster can realistically do to be prepared for all contingencies is develop a set of forecasts based on alternative weather possibilities.

The business cycle influences water use in more subtle ways, which depend heavily on the characteristics of the utility's customers. If most customers are residential and commercial, the primary impact of the business cycle is that some people's incomes will be lower than normal. These people may respond by cutting water use (among other things). If the utility has industrial customers using process or cooling water, their demand may be a function of production schedules, which result from product sales. A slowdown in the local economy or the national or international market for the product of local firms may result in large declines in water use.

Business cycles have been studied intensively for over a century, but economists have not been able to forecast either the timing or severity of the cycles. This poses a problem for utility forecasters, since population and income growth, as well as local commercial/industrial activity, depend to some extent on the local business cycle. Many university research departments produce state and local economic forecasts using models of the local economy combined with forecasts from national organizations. The water demand forecaster can use these economic forecasts as independent variables in a causal/structural model.

Causal/structural models based on grouped customer data often provide good estimates of sectoral water use in the medium-term forecast horizon. Such models also provide a valuable framework for thinking about the future. However, reliable forecasts of independent variables may be as difficult to produce as the forecast of the variable of interest. One approach to resolving this dilemma is to develop alternative scenarios with values of the independent variables consistent with high, medium, and low future water use. This method not only produces a range of possible values for the forecast, but focuses attention on the characteristics of the local economy that would produce each result.

Forecast Interval Estimates

Forecasts generally gain credibility by offering an interval rather than a point forecast. Thus a weather forecast stating that the low tonight will be between 45° and 48°F is more believable than a point forecast for 47°F. There are several approaches that are readily available to generate forecast intervals, including those that follow.

Regression-Generated Confidence Intervals

Regression analysis and other statistical approaches to forecasting provide methods for computing the confidence intervals of forecasts. The general result is that the distance from the lower to the upper bound on a 95 percent confidence interval is related to

- the standard error of the regression
- the difference(s) between the mean value(s) of the explanatory variable(s) and the projected value(s) of the explanatory variable(s)

If one explanatory variable, such as population, is projected to grow or decrease (i.e., exhibit a trend), the confidence intervals will grow larger and larger over time.[1]

In arriving at an overall confidence interval for a forecast, it also is necessary to take into account the forecast errors for the explanatory variable.

Monte Carlo Simulations

Often, the best way to take all of the possible variations in a forecast into account is to conduct what is called a Monte Carlo simulation, or risk analysis (refer to Jones 1991).

Simple Monte Carlo simulations can be carried out with standard business spreadsheets. As an example, suppose the object is to assess a realistic confidence interval for a per capita water demand forecast in year N, when population projections are uncertain. One approach to this problem is to establish the high and low population forecast for year N and randomly pick a number between this upper and lower bound to represent the actual population in that year. Similarly, the per capita water use is randomly selected from the regression confidence interval applying to year N. Combining these randomly drawn values for population and per capita water use produces one of a range of possible water use estimates for year N. Repeating this process 1,000 or 10,000 times produces a frequency distribution for estimated water use in year N, reflecting the assessments of the probabilities of the contributing factors.

Alternative Scenarios

Alternative values of the independent variables of the forecasting model can be developed from a set of scenarios of the future. Thus, a medium- to long-term demand forecast might include a low-growth scenario, in which the economy and population of the utility area grow slowly, if at all; a high-growth scenario driven by new industries and a revitalized infrastructure; and a medium-growth scenario. Each scenario would be used to develop a consistent set of predictions for values of the independent variables, which would then be incorporated into the formal model to produce three alternative forecasts. The scenario discussions would elucidate for the forecast user the underlying forecast assumptions, thereby making the forecast more valuable.

Forecast Evaluation

Forecast evaluation starts with computation of the forecast error, which is the difference between the actual values of the dependent variable and the forecast values. Several of the most widely used measures of forecast accuracy are presented here.*

Mean Squared Error (MSE)

The differences between the predicted and actual values are squared and added together, creating a sum of squared errors that would be 0 if the forecasts were perfectly accurate. The sum of squared errors accords the same weight to positive and negative forecast errors, while weighing one large error much more heavily than several small errors adding up to the same total. The sum of squared errors is typically divided by N, the

*Other error measures may be found in forecasting texts, such as Hanke and Reitsch (1989) and Armstrong (1985).

number of observations. This yields the MSE, which can be compared among samples of different sizes. Using monthly data for the example, the MSE is

$$MSE = \frac{1}{N}\sum_{i=1}^{N}(Q_i - \hat{Q}_i)^2 \qquad (11\text{-}1)$$

Where:

N = the total number of months in the sample

i = the sequence number of each individual month

Q_i = the observed water use during the ith month

$\hat{Q}_i$ = the estimated water use during the ith month

The square root of the MSE is the standard error of the estimate, which provides one measure of the "typical" size of the forecast error.

Mean Absolute Deviation (MAD)

The MAD is an alternative measure that is used when the analyst desires to weigh one large error the same as several small errors adding up to the same total. The MAD considers positive and negative errors equally by taking their absolute values. Using the same variables as Eq 11-1, it is

$$MAD = \frac{1}{N}\sum_{i=1}^{N}\left|Q_i - \hat{Q}_i\right| \qquad (11\text{-}2)$$

The MAD provides a measure of the typical, or average, size of the error or difference between the actual data and the forecast. The most important difference between the MSE and the MAD is that the MSE squares the individual errors, thus putting heavy weight on outliers, while the MAD does not.

Mean Absolute Percent Error (MAPE)

The forecast error is often presented in percentage terms by dividing each measured error by the actual value of its corresponding observation. This provides an error measure that is unit free, so that the accuracy of forecasts can be compared for different-sized utilities or market segments. The MAPE is

$$MAPE = \frac{1}{N}\sum_{i=1}^{N}\left|\frac{Q_i - \hat{Q}_i}{Q_i}\right| \qquad (11\text{-}3)$$

Bias

The bias of a forecast is its tendency to either exceed or fall short of the actual data values. A widely used measure of bias is the mean error

$$\text{Mean error} = \frac{1}{N}\sum_{i=1}^{N}(Q_i - \hat{Q}_i) \qquad (11\text{-}4)$$

The mean error would be 0 if the sum of all overestimated values exactly offset the sum of all underestimated values. The mean error can also be put in percentage terms, giving the mean percent error, or the average percentage by which the forecast strays from an even balance of positive and negative errors. Its value is

$$\text{Mean percent error} = \frac{1}{N}\sum_{i=1}^{N}\frac{(Q_i - \hat{Q})}{Q_i} \qquad (11\text{-}5)$$

Error Measures and Loss Functions

An analyst will probably compute several measures of forecast errors in carrying out a forecast evaluation. The error measure of most importance depends on the losses associated with various types of error. If one large error is more important than several combined smaller errors, a measure such as the MSE, which uses a sum of squared errors, is most appropriate. For example, the political cost to water utility officials may increase more than proportionately with the size of a water shortage.

If instead the loss imposed by a forecast error is simply proportional to its size, a measure such as MAD is appropriate. In some situations, the loss may be more closely associated with the relative error size, so that one of the measures using a percentage would be appropriate.

There is no particular reason to believe that the loss imposed by forecast errors is symmetrical. For example, the political/economic cost of a 20 percent supply shortage may be much more severe than that of an equal size surplus or excess capacity. In this case, positive errors (actual greater than forecast) would be given a much greater weight than negative errors (actual less than forecast). For a revenue forecast, the opposite may be true, so that a $1 million deficit is much more serious than an equal surplus.

Strategies to Improve Accuracy

The most important source of improved forecast accuracy accrues from a careful evaluation of past forecast results. Evaluation will reveal the size and pattern of forecast errors, and often will suggest ways to improve the next set of forecasts. The greatest obstacle to improvement is a failure to compare prior forecasts with actual events.

In a regression-based forecast, the analyst can disaggregate the total forecast error between errors in the values of the independent variables and error resulting from the model specification. A retroactive forecast using the actual values of the independent variables should be produced in order to sort out these effects. The realized value of each independent variable can be compared to its forecast value to determine errors in forecasting the independent variable values. The forecast may have been off primarily because the weather was abnormal. In this case, no adjustment in the forecast model is warranted. Conversely, the error may have arisen from the omission of some important variable from the model specification, or from a change in the utility system. These determinations would lead to a reformulation of the model.

Forecast error could also have arisen from errors in the historic data used to estimate the model parameters. Efforts to improve the available data can be highly beneficial.

When uncertainty about the general characteristics of the future is great, forecast accuracy can often be improved by averaging alternative forecasts of water use. Instead of putting all of his or her effort into a single forecast model, the analyst could use the same effort to develop three or four models using a wide range of techniques. The results of these models would then be averaged to obtain a "best guess" forecast. At least one national forecasting service follows this strategy by obtaining forecasts of inflation, interest rates, economic growth, etc., from several dozen individual forecasting organizations, and then takes an average to obtain a consensus forecast.

Before committing additional resources to a forecasting effort to improve accuracy, the forecaster and the forecast user need to jointly consider the costs and benefits of forecast improvement. This requires an evaluation of the expected incremental benefit to the organization from forecast improvement compared to the incremental cost of that effort. Additional resources would only be devoted to the forecasting effort when expected benefits exceed expected cost.

Why Forecast Formally?

Taking all the potential sources of variation in a forecast of water use for an urban area into account may seem to lead to indeterminate answers. However, this is not necessarily true, since some risks tend to balance out. Indeed, as Spyros Makridakis states in the following quote (Makridakis 1986), it may be in the medium-term forecasting period where risks are amplified to their greatest extent:

> In the long term, economic laws and S-shaped curves always prevail, making it possible to identify and measure long term trends and average relationships using least squares methods .... In the short term, seasonality rarely changes. In addition, the momentum or inertia of most phenomena makes extrapolation routine and allows the use of simple methods (such as exponential smoothing and decomposition) which isolate and project accurately the most recent pattern. The combination of seasonality and inertia/momentum is indeed being used to a great extent by most statistical methods to make their predictions which on average are more accurate than ... judgment. However, forecasts must be made for additional horizons including the medium term, which is the hardest to deal with because the depth and length of each cycle is unique .... Only through a combination of judgment and systematic analysis of past fluctuations can we arrive at reasonable conclusions about medium term cyclical fluctuations which can help us more accurately monitor and more successfully predict future cycles and the resulting uncertainty.

The question is not "Can we develop a perfect forecast?" The operational question is "Can the utility's demand and revenue forecasts be improved by developing a formal forecasting program?" Even without much data, and using only simple models, a utility can develop forecasts of water demand that will be more accurate than the implicit forecasts used in the absence of a formal effort.

Forecasting is a continuing project. The initial forecasting model should evolve and become more accurate over time. Accumulating experience, including a careful comparison of forecasts and actualizations, can suggest refinements. When conditions are changing slowly, and water rates are incrementally increased each year, there may be little apparent need for a comprehensive forecasting model. However, for contingency planning, the incorporation of weather variables, population, price, and conservation effects is necessary to obtain accurate revenue predictions under various unusual circumstances, and therefore aid in preparations to handle such circumstances ahead of time.

There is no guarantee that the future can be accurately predicted. Nevertheless, it is usually feasible to analyze water use patterns, and to obtain relatively accurate forecasting models for next month's water use or for periods of up to several months. It is also possible to develop information for long-term planning efforts with regression models based on such explanatory variables as population, income, price, etc. These really are hypothetical models in which the random element that is so apparent in demand on a day-to-day basis is factored out over the long run. Beyond that, to make credible statements about water demand in an urban system over a period of several years requires careful study of the available data and general knowledge about population, climate, socioeconomic, and other relevant processes.

To develop forecasts that are as accurate as possible, the forecasting function should be independent of the planning and operating activities of the utility. It is imperative that the forecast not be unduly influenced by what management wants to hear. The forecast will take utility operating policies and programs into account, but the forecaster should have the independence to report results that disagree with management expectations.

Notes

1 Suppose $y = \beta x + \varepsilon$ expresses a simple regression of y = total system water use with x = population as the independent variable. Both x and y are expressed as deviations from their respective means. Then, the variance of the forecast error, when x^* is a projected population for a given year, is

$$Var\,(y^*) = S^2 \left(1 + \frac{(x^*)^2}{\sum\limits_{sample} x_t^2} \right)$$

where x_t is summed over the sample or historic data on population and water use. Thus, the larger x^* is, i.e., the forecast value of population is from the average population figure in the sample or historic data, the larger the variance of the forecast and the larger the confidence interval. Here, S is the standard error of the regression. The formula for the more general case involving a multivariate regression with several explanatory variables is provided in note 3 of chapter 5 (page 68).

References

Armstrong, J.S. 1985. *Long-Range Forecasting: From Crystal Ball to Computer*. New York, N.Y.: John Wiley & Sons.

Hanke, J.E., and A.G. Reitsch. 1989. *Business Forecasting*. 3rd ed. Boston, Mass.: Allyn and Bacon.

Jones, C.V. 1991. *Financial Risk Analysis of Infrastructure Investment*. Greenwood, Conn.: Quorum Books.

Makridakis, S. 1986. The Art and Science of Forecasting: An Assessment and Future Directions. *International Journal of Forecasting*.

Appendix A

Direct Data Collection

Water utilities often want to gather information about their customers beyond the basic water use data generated by the accounting system. While some demographic information is available from outside sources, such as the decennial census of population and local governmental agencies (discussed in chapter 2), more detailed information can only be obtained directly from the utility's customers. Desired information might include customer response to publicity about water conservation, the rate at which they accept water-saving devices, their willingness to spend for water system improvements, and the extent to which they might be willing to pay now to avoid some probability of water rationing in the future.

This appendix touches on some considerations that must be taken into account when designing and conducting surveys. Topics discussed include methods of customer feedback, development of alternative scenarios, random sampling, development of questionnaires and questions, selecting a sample, determining the sample size, and a discussion of mail and telephone surveys.

Customer Feedback

One low-cost source of information is provided by day-to-day customer contact with the utility. While most customers simply pay their bills, a small percent will spend time discussing water-related problems with utility employees. While this feedback does not provide a statistical basis for inferring how the vast majority of customers feel, a careful tabulation of these comments and complaints provides continuous feedback on how well the utility is serving customer needs. Additional low-cost information may be obtained from such sources as letters to the editor of local newspapers and the results of water-related bond elections.

Opinions about the future can also be collected from current and prospective water customers. For example, major industrial water users could be asked to share their plans for changes in their technology and output, as well as the resulting impact on their water demands. Similar information could be obtained from governmental regulators, who may be developing new regulations on the use of water or of disposal of wastewater, either of which would impact water demand over the next few years. Survey results could be directly compiled into a forecast of future water demands, or, more likely, the survey results would be incorporated into the forecast either as a judgmental adjustment or as values for an independent variable in a causal/structural model.

Expert Opinion

Alternative Scenarios

One approach used to examine possible future trends involves the development of alternative scenarios. For the long term of a decade or more, scenario analysis seems advisable as a complement to a complex, multivariable, causal/structural model of demand. This is because each of its several explanatory variables must themselves be forecast. Qualitative issues may be more important in the longer run, especially those pertaining to directions in the local or regional economy. For many areas, the key questions are whether industrial decline will be reversed or vigorous economic growth reinstituted. For other areas, the problem is whether or not the current soundness of the economy will be sustained in the face of challenges from global competitors. Scenario analysis is useful in this regard. This approach involves brainstorming about how future trends will play themselves out for a particular urban area under several alternative assumptions.

Delphi and Other Expert Opinion Techniques

Intuitive forecasts take a number of specific forms, of which the Delphi methods are the best known. The advantage of such forecasts is the incorporation of expert opinion—without the time and cost of formal model building. The process is not necessarily cheap, however, since considerable time and effort are required to obtain the desired responses, and it may be necessary to pay the experts.

The Delphi method for forecasting water demand would use a group of experts selected from a wide range of specialties, such as civil engineering, sociology, economics, economic development, land development, urban agriculture, and political science. The experts would each be asked a series of questions about the long-term future, such as the following:

- What will happen to exterior water use by each major category of customer?
- What will happen to indoor water use by each major category of customer?
- How much will the size of each customer class change?

Each expert writes out a set of responses and submits it to the facilitator, who prepares anonymous summaries of the various opinions and sends them back to the panel members. Each panel member then has an opportunity to modify his or her forecast in light of the opinions of other experts. This process goes through several iterations, with an end result that will either lead toward consensus or to a clarification of why different experts hold varying opinions on the most likely outcome.

Expert panels with personal interaction can also be used. In this method, the experts assemble and share their viewpoints with each other. This has the advantage of allowing direct interaction, which may lead to quicker clarification of differences or movement toward a consensus. The primary disadvantage is that the more vocal or forceful members of the panel may dominate the discussion, leading to less accurate forecasts. This process also leaves little time for reflection, which may short-circuit the development of new insights.

Regardless of the process used, the experts making up the panel should be diverse in training and experience, and should include at least some members with no stake in the results of the forecast. A wide diversity among the panel members helps to minimize the bias that tends to develop among any group in the same geographic or professional area.

Various attempts have been made to assess the accuracy of judgmental forecasts. Armstrong (1985) concluded that for long-range forecasts, causal/structural models are superior to judgmental models.

Designing a Survey

Although valuable information can be obtained from a variety of sources, only a well-designed effort using statistically based random sampling models is capable of providing data from which valid inferences about the entire population can be drawn.

When a utility decides that information about customers is useful and worth the required resources, several steps, some of which may require the assistance of a professional statistician, are necessary in order to carry out scientifically valid research. (See Cochran [1963], for example.)

Steps in Designing a Survey

1. Determine the question or problem requiring a solution. There is no point in spending resources to collect information unless that information can be used to help solve a problem. Problems that customer information might help solve include the need to conserve water, the need to plan for expansion of capacity, or to determine customer preferences between current water rate financing and bond financing of utility capital expenditures.

2. Determine what information will help solve this problem and how the information will be used once it is collected. There is no value in collecting information unless there is a plan showing how it will be used. Carefully thinking about how the new information will be used is essential in designing questions that will yield the desired information.

3. Determine the population about which more information is needed. For a water utility, this may be all customers, only residential or commercial customers, only customers in certain parts of the service area, or only customers with certain equipment at their homes or place of business. Additionally, the population

of interest may be only those customers whose water use exceeds a certain level, either continually or once in a while.

4. Determine how much precision is needed. While a substantial part of the cost of a survey results from developing appropriate questions and testing them, another significant part of the total cost results from the number of customers in the sample. Increased precision carries a higher price tag.

In this context, precision measures the closeness of results of repeated samples of the same size, and can be estimated before collecting the data based on statistical theory. For example, sample results are often said to be ± 5 percent except for a 1 in 20 chance. This simply means that if we collected a large number of samples from the population, the mean would vary by no more than 5 percent from the true value for 95 percent of the repeated samples.

The utility can opt for a low-budget effort with correspondingly low precision, or high precision at a much higher cost. If statistically valid results for various subgroups of the population are required, the total sample must be commensurately larger. Sample size will be discussed in detail later in this appendix.

One important limitation to the accuracy of the survey results is that respondents may systematically misunderstand or misrepresent their situation. It is well known, for example, that survey respondents consistently overstate the extent of their recycling activities while understating their consumption of tobacco products and alcoholic beverages. This source of error can be reduced but not eliminated by careful question design. It can also be compensated for by use of other sources of data.

Questionnaire Development

Developing an appropriate set of questions is one of the critical phases of obtaining information from customers. Each question needs to be clearly stated so that every respondent will understand what is being asked. The questions must avoid even subtle bias or the impression of a preferred or "socially correct" response by the customer. This is sometimes difficult, but it is well known that respondents unconsciously want to please the interviewer by giving the "correct" answer. Thus people tend to claim to recycle their beverage containers and newsprint much more than they actually do, and will claim greater interest in water conservation than they actually have. It is important, then, that questions be specific and nonjudgmental. Often this can be more effectively accomplished by using questions with multiple-choice responses. For example, to determine a customer's knowledge about water rates, the question could be asked, "Do water rates (per 1,000 gallons) vary during the year?" Answers might include: "(1) Yes, they are higher in the summer; (2) Yes, they are higher in the winter; (3) No, they are the same all year; or (4) I don't know."

The questions should be sequenced to avoid having questions near the beginning that might bias the answers near the end. The questions also need to progress from easy to difficult, to build the confidence of the respondent in his or her ability to answer the questions. A progression from general to personal is also helpful, as the respondent's commitment to completing the interview hopefully increases. The questionnaire should also avoid being too long or potential respondents will lose interest and skip some questions or not send in the responses at all.

Reviewing Questions

Subjecting the draft questions to rigorous review is extremely important. The first review should be by water utility officials and other professionals, followed by a test sample of water customers. Pretesting often reveals unexpected problems with the questions and provides an opportunity to avoid serious error.

Regardless of what medium is used for the survey, obtaining voluntary participation from the selected sample requires careful planning. The interest of the potential respondent needs to be quickly piqued. The introduction to the interview process should demonstrate the interviewer's credibility and make clear the importance of the information to be provided. The interviewee must be confident that he or she is not trying to be sold something. The attention and cooperation of the interviewee can be obtained with simple questions that can be readily answered before asking difficult questions or questions eliciting personal information such as the respondent's income group.

Target Population

Since the population of interest will usually be represented by either water meters or water customers, it is relatively simple for the utility to specify a target population and to determine the individuals in that population, along with their mailing addresses.

If, however, the population is, for example, apartment complexes, an enumeration of all water meters assigned to this category will likely include some properties that have been misclassified, either because of recording error or change in use. It is also likely that an enumeration will miss some apartments that were incorrectly classified under other headings. These random errors may not significantly alter the results, however.

Also adding to the problem is that some complexes will have several meters and others none, having been hooked up without benefit of accompanying paperwork. In some complexes, individual apartments may be metered. Common areas frequently have water service separate from the residential uses of the tenants.

The problem of identifying the target population becomes substantially more challenging when the utility is interested in customers with certain characteristics not observable within the utility accounting system. Such characteristics may include the ownership of a swimming pool or the use of water filters for household drinking water. The presence of permanent swimming pools could be determined from aerial photography or from property assessor's records, while use of water filters could be determined only by asking the individuals within the context of a more general survey.

Nonresponse Bias

Another important limitation to the precision of a survey's results is nonresponse by some customers to the survey efforts. Nonrespondents' answers may be quite different from those provided by the group answering the questions.

Efforts to increase the response rate dramatically increase the cost of a survey. To the extent that the characteristics being investigated are correlated with water use per customer, the bias in the results may be less important than for other surveys, since the utility knows the water use of each respondent and nonrespondent. Note that bias in the results caused by nonresponse would still be a problem even if the utility attempted to obtain information from the entire population. Increasing the sample size has no effect on this source of error.

Selecting the Sample

The simplest sample selection method is a random sample of the entire population to determine some characteristic of that population, such as the proportion of homes having low-flow showerheads. In this case, the entire population is listed (perhaps as a list of all single-family residential water customers in the utility computer), and a table of random numbers is used to select a sample of the desired size. Typically, the computer will have a routine to generate these random numbers and make the sample selection.

Systematic Sampling

Frequently, the task of drawing a sample can be simplified by substituting a systematic sample for a purely random sample. In this case we might select every 20th water customer from the utility list of customers. The selection would be started by drawing a random number between 1 and 20. This eliminates placing all of the customers in a numeric sequence and drawing random numbers to determine which ones are in the sample.

A systematic sample will approximate the results of a truly random sample as long as the individual items are randomly distributed within the overall file with respect to the variables of interest. This would generally be valid for customers arranged alphabetically by name or by arbitrarily assigned billing numbers. Selecting every 10th house along each street, however, could result in systematically including too many corner lots or excluding most corner lots.

Sample Size for a Proportion

The sample size selected for a survey depends on the precision desired and, surprisingly, not on the size of the population, unless the population is relatively small (under 10,000). Sample size is also based on estimates of the standard deviation of the variable being studied, such as water use. The basic relationship is that the variability of the sample mean equals σ/N, where σ is the standard deviation of water use per customer in the population of the customer class being considered. Furthermore, the sample mean is an unbiased estimator of the population mean (regardless of the size of the population), and the sampling distribution tends toward a normal distribution as the sample sizes become larger. All this means that it is possible to derive simple formulas linking the desired tolerance of the estimate to be derived from the sample with the variability of the phenomena under consideration.

As an example, if the proportion of households with low-flow showerheads is to be determined with an error of $\pm$ 5 percent, and we are willing to accept a 1 in 20 chance of being outside this range, the sample size would be no more than 400. Required sample size for a proportion is

$$N = PQ\left(\frac{t}{d}\right)^2 = (0.3)\,(0.7)\left(\frac{1.96}{0.05}\right)^2 \qquad \text{(A-1)}$$

Where:

P = the proportion of the population expected to have low-flow showerheads

$Q = 1 - P$

t = the statistic from the normal distribution indicating the chance of a greater than chosen error (1 in 20 in this example)

d = the acceptable proportion of error (5 percentage points in this example)

Thus, if 30 percent of the population is expected to have low-flow showerheads, the required N computed in Eq A-1 is 323. If a response rate of 40 percent is expected, 808 questionnaires would be sent out. If the total population being sampled was smaller than 10,000, the sample size could be reduced. The formula for reducing sample size for small populations, known as the finite sample correction (FSC) is

$$FSC = \sqrt{\frac{population - sample\ size}{population - 1}} \tag{A-2}$$

which reduces the estimate of N slightly. For example, if the total population were 4,000, the finite sample size correction factor is 0.96, which would reduce the sample size from 323 to 310.

If separate results are required for various groups within the total population, several options exist.

If it can be determined which customers are in each group before doing the survey, a stratified random sample can be used, and the sample size within each group can be determined in the same way as for a single population. Thus, if the utility wishes to obtain results by group of customers, and these groups can be identified through the utility's computerized records, then the task is straightforward. For example, single-family residential customers might be divided into low-, medium-, and high-volume water users. Each customer would be placed into the appropriate category using utility billing records, and a random sample would be drawn from each category. Alternatively, the sample could be stratified by census block or tract (perhaps based on income levels), postal zip code, or political district.

When customers will be placed into subcategories based on their survey responses, it is harder to determine the needed sample size. In this case, the analyst needs to make some initial guess or estimate of the proportion of the population in each subcategory, and select a large enough sample of the total population to make the results for each subcategory statistically valid. For example, suppose that the analyst plans to divide customers into those with one bathroom, two bathrooms, or three or more bathrooms. Note that this will require a definition of what fixtures are necessary to constitute a bathroom.

An estimate could probably be obtained for this distribution from real estate agents or home builders in the service area. Suppose they estimate that 20 percent of the homes have one bathroom, 68 percent have two, and 12 percent have three or more. In order to obtain a large enough random sample to ensure statistical validity for the group with three or more bathrooms, 323 responses are required from this group. This would require a simple random sample of about 2,700 responses and a mailing of 6,750, assuming a 40 percent response rate.

If the sample could be stratified ahead of time, perhaps by using property assessor's data or census block data (since home value and number of bathrooms are highly correlated), the necessity of this large sample could be avoided. Prestratification of the sample would only be done if it is less expensive than increasing the sample size sufficiently to obtain the desired accuracy for the smallest strata.

Mail Survey

It is relatively inexpensive to do a one-shot mail survey, perhaps sent as an insert in a random sample of monthly water bills meeting some specified criteria. A more intense program would reduce nonresponse, but each added step increases cost. For example, a survey program might include several or all of the following steps (Dillman 1978).

1. Mail a postcard or short letter explaining why survey information is important to the city. It may be helpful to have the message sent by a recognizable figure, for example, the mayor.
2. A few days later, mail the survey form.
3. After about 10 days, mail a reminder postcard to nonrespondents.
4. After another 10 days, mail a second questionnaire to those still not responding.
5. Mail a second reminder postcard after another 10 days.
6. Mail a certified letter with another copy of the questionnaire to a sample of remaining nonrespondents after another two weeks.
7. Send a trained interviewer to visit a sample of remaining nonrespondents after another two weeks.

The reason for intensively pursuing information from only a sample of nonrespondents is to hold down the total cost while still determining the extent to which previous nonrespondents differ from those who initially responded.

Few survey efforts would go to the extent described, unless there was reason to believe that nonrespondents were quite likely to be different with respect to the activities of interest than those who responded. Any survey that includes followup requires keeping track of who submitted each completed questionnaire and appropriate assurances of confidentiality to those answering the questions. Tracking is also essential if the survey responses are to be matched up with the utility's water use records. Survey information can also be enhanced by matching it with property assessor's records of the physical characteristics of each property. If such followup or matching is going to be used, it is necessary to directly tell the potential survey respondent and give appropriate assurances of the confidentiality of the responses. This requires follow-through to make sure the responses remain confidential, are used only for statistical analysis, and do not single out individual customers.

If the utility plans to single out individual customers based on their responses, such as for an audit of outdoor water uses, this should be clearly stated in the introduction to the survey.

Telephone Survey

While obtaining a sample of water customers and their mailing and/or service addresses is relatively simple, given the utility billing records, obtaining telephone numbers to conduct a telephone survey may be quite difficult.

For surveys of business customers, phone numbers are generally readily available; the yellow pages of the telephone directory provides a ready source of possible customers by category. For example, a utility might wish to gather information from plumbing companies regarding the water use rates of faucets, showerheads, and toilets that they install. The population of plumbing companies is approximated by the yellow page listings, and a random sample could readily be developed. As mentioned, the cost will escalate as efforts to obtain responses from nonrespondents increases. Numerous callbacks may be required to obtain information from some business owners or managers, if they are willing to respond to questions at all.

An additional cost of telephone or in-person surveys is training and monitoring the interviewers to maintain accuracy and consistency. Respondents have the opportunity to ask for clarification of various questions, and their answers may be open to interpretation by the interviewer. Utilities or others wishing to conduct telephone surveys frequently hire commercial survey firms for this reason. The survey of Tucson, Ariz., apartment managers described in chapter 10, some results of which are provided on the data diskette, was done by a professional telephone survey company. A notable advantage of this was that data was entered into a computerized database as it was collected by the interviewers, thus avoiding the later task of data entry that is encountered in a mail survey. Keep in mind that there is a considerable setup cost of entering the questionnaire into the computer system and testing it for errors. An added advantage of the computerized interview system was that the program allowed irrelevant questions to be automatically skipped over in real time. If a respondent reported no grass, then the series of questions on grass irrigation was automatically skipped, taking the interviewer to the next relevant question.

An Example Telephone Survey

To develop the sample for the Tucson apartment survey, a list of all apartment complexes served by the water utility was obtained, and then a random sample was drawn from the list. Obtaining telephone numbers was an unexpectedly difficult challenge. The telephone book, city directory, apartment locator guides, and other sources were used. Finding telephone numbers for the larger complexes proved to be quite easy, while those for small complexes were not always available. Some owners of small complexes do not list a telephone number in any of the published sources and do not have listed personal phone numbers.

While it was impossible to conduct telephone interviews with apartment owners or managers whose phone numbers were unknown, there was also a large number of nonresponses resulting from no answer or response only from an answering machine. Virtually all managers or owners who were actually reached by telephone agreed to participate in the survey. Carrying out the survey under sponsorship of a major state university may have increased the credibility of the effort and enhanced the response rate. One unexpected quirk in the survey was that some individuals managed several apartments in the sample and were interviewed more than once concerning different properties. If the project budget had been larger, an interviewer could have been sent to a sample of apartments with no listed phone number or where no response was obtained.

An Alternative Approach

An alternative approach to telephone survey sampling is to take a random sample of all telephone numbers in the survey area (excluding emergency numbers, hospital numbers, etc.), and qualify the person answering the phone. Qualify means to determine if the telephone is associated with the target population, such as single-family residential customer, etc. This requires a much larger number of total phone calls, but with automated dialing equipment, this may not be a problem. Having a real person on the phone, however, is undoubtedly more effective in obtaining responses. Few people will listen to a machine unless they really want to make or receive a specific phone call.

Summary

When a water utility decides that it needs information about its customers beyond that generated in the normal course of business, direct data collection is a highly viable option. Information about customer behavior and attitudes can be obtained directly from customers by using sampling techniques. Such data can then be used to improve utility decision making regarding conservation programs, pricing initiatives, and other policies.

Long-range planning can be aided by use of intuitive information gathering and analysis. Popular approaches include Delphi techniques, scenario building, and customer surveys. Gathering information from as wide a range of sources as possible will aid the utility decision makers in making the best possible assessment of future long-run trends. Such data can help avoid costly policy blunders that could result in (1) large excess capacity and its accompanying cost, (2) insufficient capacity that would deprive customers of their accustomed level of service, or (3) unacceptable changes in water quality.

References

Armstrong, J.S. 1985. *Long Run Forecasting, From Crystal Ball to Computer*. New York, N.Y.: Wiley Interscience.

Cochran, W.G. 1963. *Sampling Techniques*. New York, N.Y.: John Wiley & Sons.

Dillman, D.A. 1978. *Mail and Telephone Surveys—The Total Design Method*. New York, N.Y.: Wiley Interscience.

Glossary

aggregation The combining of individual observations to obtain a new set of data representing the average or total values of the original set.

ARIMA Mixed autoregressive moving-average model used for forecasting a single time series using only data from the history of that time series itself.

autocorrelation The correlation or relationship between successive observations in a time series. The correlation of the series with itself lagged one period.

average price Applied to water, the customer's total water bill divided by the quantity of water purchased.

***b* coefficient** In a multiple regression model, the estimated relationship between the dependent variable and an individual independent variable. The slope of the function between two variables.

backcast An estimate of the values of a data series for a time period prior to that used in estimating the model. Backcasts are sometimes used to test the validity of a model that will be used for forecasting.

block rate A rate structure in which the marginal price of water changes based on the quantity purchased.

Box-Jenkins model *See* ARIMA.

business cycles Recurring multiyear fluctuations in the economy, consisting of an economic expansion followed eventually by a recession or contraction. Cycles vary in length from less than two years to more than eight years.

causal/structural model A model in which variations in the variable of interest (dependent variable) are explained by variations in one or more independent variables.

Ccf Hundred cubic feet. 1 Ccf = 748 gal.

climate The long-term tendencies of temperature, precipitation, wind speed, etc., that characterize a particular region.

coefficient model A model in which the water use rate per capita (the coefficient) is multiplied by total population to forecast total future water use.

conditional forecast Uses projected values for the period being forecasted. For example, these might include expected future prices for water and personal income.

confidence interval Used to illustrate the precision of an estimate or forecast. The confidence interval might include the range in which the value of a variable is expected to occur 95 times out of 100. Thus, a 95 percent confidence interval leaves a 5 percent chance that the true value is outside the interval shown.

correlation The degree to which two data series are related. A correlation coefficient of 1 means that the two series are perfectly related (both change together in exact unison). A coefficient of 0 indicates no relationship, while a coefficient of -1 shows an exact inverse relationship.

cross-section data Data on a number of different units, such as households, census tracts, or water utilities. Generally used to explain differences among these units.

decomposition The strategy of breaking a problem into subproblems, solving them, and then recombining the individual solutions to get an overall result.

decreasing block rates A rate structure in which the price per unit falls as more is purchased.

degrees of freedom The total number of observations (such as months of data) minus the number of coefficients estimated by some statistical procedure, such as multiple regression.

Delphi A method of obtaining forecasts from a panel of experts.

demand As used in economics, the quantity of a product that will be purchased at each possible price, given that all other prices, consumers' incomes, stocks of durable equipment, habits, and preferences are held constant. Also referred to as the demand curve.

dependent variable The variable of primary interest to the analyst. The variable that is being forecasted. The variable on the left-hand side of a regression equation.

dummy variable A variable that only takes on values of 1 or 0, representing the presence or absence of a particular condition. For example, a drought dummy variable would have a value of 1 during a drought and 0 during normal years.

Durbin-Watson statistic A measure of the correlation of successive error terms from an estimated time-series equation (between errors in time t and time $t-1$). A value of 2 shows no autocorrelation is present, while values smaller than 1.5 or larger than 2.5 indicate a problem that should not be ignored.

elasticity An expression of the relationship between two variables showing the percentage change in the dependent variable for a 1 percent change in the independent variable. For example, the price elasticity of demand shows the percent change in water use for a 1 percent change in water price.

end-use model A forecasting model developed by adding up the end-uses for water, such as toilet flushes, showers, lawn irrigation, etc.

endogenous variable *See* dependent variable.

error term The difference between the actual value of a variable and its estimated value from a statistical model.

evapotranspiration The water loss from leaf surfaces and soil. Used to measure irrigation requirements.

exogenous variable *See* independent variable.

explanatory variable *See* independent variable.

extrapolation models Forecasting models in which past variations in the dependent variable are used to estimate future values.

exponential smoothing A univariate forecasting model that removes the irregularities from an historic time series in order to create a forecast. This is a special case of the more general ARIMA models. Used for short-term forecasts.

***F* statistic** In regression, used to describe the statistical significance of the entire model.

forecasting Estimating future values of a dependent variable in "unknown" situations.

forecasting model A model used to obtain forecasts.

goodness of fit How well the values estimated by a model fit the actual, known data values. There are a number of different statistical measures used to assess the goodness of fit.

heteroscedasticity The general size of the error term from an estimated model varies over the range of data. The error may be small when the size of the variables in the model is small and large when they are large.

homoscedasticity The variability of the error term is constant over the entire range of data.

increasing block rates The price per unit increases as the quantity purchased increases.

independent variable A variable used to help explain variations in a dependent variable. A change in an independent variable is said to cause a corresponding change in the dependent variable. A causal variable.

inflation A reduction in the value of money, measured as the percentage increase in the weighted market prices of a basket of consumer goods.

interaction A situation in which the relationship between X and Y depends on the level of X. For example, the price elasticity of demand for water may increase as the price of water increases.

intercept dummy variable *See* dummy variable.

long-term forecast A forecast for time periods generally 10 or more years ahead.

log-normal variable The logarithm of the variable is normally distributed.

logistic curve A growth curve in which the percentage growth gets smaller over time. Also known as the S-curve. It can be computed from $1/Y = 1/a + cX^t$, where Y is the dependent variable, a is the asymptote, c is a coefficient to be estimated, X is the growth factor, and t represents time.

MAD Mean absolute deviation. The average of the absolute values of the error terms of a forecast or estimate.

MAPE Mean absolute percent error. The average percentage error of a forecast or estimate, ignoring the sign of individual errors.

marginal price The price of the last unit purchased. It may be (1) lower than average price when a water rate schedule includes decreasing block rates, or (2) higher when it includes increasing block rates.

mean Arithmetic average.
mean absolute error (deviation) *See* MAD.
mean absolute percent error *See* MAPE.
MSE Mean square error. The average of the squared errors of a forecast or estimate. Squaring puts more weight on larger errors than the otherwise similar mean absolute deviation.
measurement error The extent to which errors result from shortcomings in the way that numbers are assigned to a concept. For example, faulty water meters or erroneous meter reading would result in errors in water use data.
measurement model A model used to estimate relationships among variables. These relationships can then be used in subsequent forecasting models.
median The observation in the center of a distribution. Half of the values in the series are larger and half are smaller.
medium-range forecast Generally includes a time period ranging from a few years up to 10 years.
microdata Data on individual customers. The opposite of aggregated data.
mode The category with the most observations.
model A representation of the most important characteristics of the real world for the purpose at hand.
model specification The choice of the set of independent variables and their form to be used in a model.
multicollinearity The extent to which two or more independent variables are correlated or vary together. High multicollinearity among independent variables makes it difficult or impossible to separate the individual effects of the variables.

normal distribution The bell-shaped frequency distribution for a variable. This distribution is symmetric around its mean value, which is also the median.

observation Measurements for a given unit (individual customer or water utility) for a given time period, such as a day or month.
outlier An observation on a variable that is unexpectedly large or small compared to the remaining observations.

parameter The estimated relationship or descriptive statistic from a computation.
peak demand The greatest amount of water delivered for a particular time interval. For example, the peak hour for a system might be from 5:00 to 6:00 p.m. on July 23.
price elasticity of demand The relative responsiveness of quantity demanded to a change in price. The percentage change in quantity caused by a 1 percent change in price.
price elasticity of revenue The relative responsiveness of revenue to a change in price. The percentage change in revenue resulting from a 1 percent change in price. Equal to 1 plus the price elasticity of demand.
principle of composition The idea that forecast accuracy is improved by disaggregation with separate forecast models for each category.
principle of noise The idea that it is easier to predict an average characteristic of a group than the specific characteristic of one selected individual—the law of large numbers.

R^2 The coefficient of determination of a regression showing the proportion of variation of the dependent variable accounted for by variations in the independent variables.

regression A statistical method of estimating the relationship between several independent variables and a dependent variable that is being explained.

reliability The extent to which a repetition of the measurement will yield the same result.

seasonality The recurring variation in temperature, rainfall, and water use (or any other variable of interest) over the year.

scenario A cohesive story about future events to be used in developing a specific forecast of the variable of interest.

sensitivity analysis The values of one or more independent variables are changed in order to observe the effect on the dependent variable.

serial correlation The correlation of a data series with itself lagged one or more periods. Measures the extent to which larger (or smaller) than average values are followed by more larger (or smaller) than average values.

short-term forecast A forecast for a period ranging from a few hours or days to a few years.

simultaneous model When the dependent variable in one equation appears as an independent variable in another equation. For example, under block rate pricing, the marginal price of water depends on the quantity taken, while the quantity also depends on the price.

slope dummy variable An ordinary continuous independent variable multiplied by a dummy variable so that the result is a variable that is either 0 or its normal value.

specification of the model The list of independent variables used to explain variations in the dependent variable.

specification error The omission of one or more important independent variables from a model or the inclusion of one or more irrelevant variables.

standard deviation A statistical measure of the dispersion of the values of a variable.

standard error of an estimated coefficient A statistical measure showing the significance of an estimated coefficient.

standard error of an estimated equation A statistical measure of the "typical" difference between the true value and the estimated value of a data series.

statistical significance The degree to which a result is likely to have occurred by chance. For example, a variable that is significantly greater than 0 at the 0.10 level incurs a 10 percent chance that the true value is 0 or negative in value.

***t* statistic** A statistic showing the significance of the coefficient for the relationship between two variables. The *t* statistic in regression is equal to an estimated coefficient divided by its standard error. As a rule of thumb, the *t* statistic should be greater than 2 (absolute value) for a useful relationship.

time-series data Observations on a variable are collected over successive time intervals, such as a series showing monthly water use for a customer or a water utility.

transformation An arithmetic operation performed on a variable, such as taking the logarithm or reciprocal of each observation.

trend The extent to which a time series (such as annual water sales) tends to systematically increase or decrease over time.

uncertainty The lack of confidence associated with a forecast.

unconditional forecast A forecast made without using any data from the time period of the forecast. Based only on historic data.

univariate model A forecasting model using historic variations in the variable of interest to estimate its future values. Contains only one variable.

validity Truth. How close to the truth is the estimate or forecast.

variance A measure of variation that uses the square of the deviation from the sample mean. Due to squaring these differences, this measure places a heavier weight on larger deviations.

Additional Sources of Information

Agthe, D.E, and R.B. Billings. 1987. Increasing Block Rates for Water: Equity, Price Elasticity, and Household Income. *The American Journal of Economics and Sociology*, 46(3):273-286.

Agthe, D.E., and K. Raffiee. 1984. Pricing as a Conservative Practice of Nevada Water Companies. *Nevada Review of Business and Economics*, 3(3):19-23.

Agthe, D.E., and C. Tinney. 1989. Production, Cost, and Increased Salinity in Water. *Journal of Water Resources Planning and Management*, 115(1):824-828.

Bails, D.G., and L.C. Peppers. 1993. *Business Fluctuations: Forecasting Techniques and Applications*. 2nd ed. Englewood Cliffs, N.J.: Prentice Hall.

Billings, R.B. 1982. Specification of Block Rate Price Variables in Demand Models. *Land Economics*, 58:386-394.

Billings, R.B., and W.M. Day. 1989. Demand Management Factors in Residential Water Use: The Southern Arizona Experience. *Journal AWWA*, 81(3):58-64.

Billings, R.B., and M. Longstreth. 1988. Unpublished survey.

——. 1989. *Project Report: Water Use and Conservation in Multiple Family Dwellings in Tucson, Arizona.*

Boland, J.J. 1978. Forecasting the Demand for Urban Water. *Municipal Water Systems: The Challenge for Urban Resource Management.*

Brown and Caldwell. 1984. *Residential Water Conservation Projects—Summary Report*. Prepared for the US Department of Housing and Urban Development.

Currie, R.G. 1981. Evidence of 18.6 Year Signal in Temperature and Drought Conditions in North America Since AD 1800. *Journal of Geophysical Research*, 86:11,055-11,064.

Darr, P., S.L. Feldman, and C. Kamen. 1976. The Demand for Urban Water. *Studies in Applied Regional Science*, Vol. 6. Martinus Nijhoff Social Sciences Division, Leiden.

Delorenzo, T. 1982. Utility Profits, Fiscal Illusion, and Local Public Expenditures. *Public Choice*, 38:243-252.

Deno, K.T., and S.L. Mehay. 1988. Municipal Utilities and Local Public Finance: A Simultaneous Model. *Public Choice*, 57:201-212.

DeRooy, J. 1974. Price Responsiveness of the Industrial Demand for Water. *Water Resources Research*, 10(1):403–406.

Ferguson, B.K. 1987. Water Conservation Methods in Urban Landscape Irrigation: An Exploratory Overview. *Water Resources Bulletin*, 23(1):147–152.

Foster, H.S., and B.R. Beattie. 1981. On the Specification of Price in Studies of Consumer Block Rate Price Scheduling. *Land Economics*, 57(4):257–265.

Griffin, A.H., and W.E. Martin. 1981. Price Elasticities for Water: A Case of Increasing Block Rates: Comment. *Land Economics*, 57(4):266–275.

Hanke, S.H., and R.K. Davis. 1973. Potential for Marginal Cost Pricing in Water Resource Management. *Water Resources Research*, 9(4):808–825.

Hansen, R.D., and R. Narayanan. 1981. A Monthly Time Series Model of Municipal Water Demand. *Water Resources Bulletin*, 17(4):578–585.

Linaweaver, F.P. Jr., J.C. Beebe, and F.A. Skrivan. 1966. *Data Report of the Residential Water Use Research Project*. Baltimore, Md.: Johns Hopkins University, Department of Environmental Engineering.

Longstreth, M., and R.B. Billings. 1990. Determinants of the Use of Low-Flow Fixtures in Apartments in an Arid Climate. In *Proceedings of the American Council on Consumer Interests.*

Martin, W.E., and J.F. Thomas. 1986. Policy Relevance in Studies of Urban Residential Water Demand. *Water Resources Research*, 22(13):1,735–1,741.

Martin, W.E., H.M. Ingram, N.K. Laney, and A.H. Griffin. 1984. *Saving Water in a Desert City*. Washington, D.C.: Resources for the Future, Inc.

Morgan, W.D., and P. Pelosi. 1980. The Effects of Water Conservation Kits on Water Use. *Journal AWWA*, 72(3):131–133.

Morgan, W.D., and J.C. Smolen. 1976. Climatic Indicators in the Estimation of Municipal Water Demand. *Water Resources Bulletin*, 12(3):511–518.

Narayanan, R., D.T. Larson, and T.C. Hughes. 1985. Effectiveness of Drought Policies for Municipal Water Management. *Water Resources Bulletin*, 21(3):407–416.

Nelson, J.O. 1979. Northern California Rationing Lessons. In *Proceedings of the Conference on Water Conservation Needs and Implementing Strategies*. Rindge, N.H.: Franklin Pierce College.

——. 1987. Water Conserving Landscapes Show Impressive Savings. *Journal AWWA*, 77(3):35–45.

Neufeld, J.L., and J.M. Watts. 1981. Inverted Block Rates or Lifeline Rates and Micro Efficiency in the Consumption of Electricity. *Energy Economics*, 3(2):113–121.

Nordin, J.A. 1976. A Proposed Modification of Taylor's Demand Analysis. *Bell Journal of Economics*, 6(2):719–721.

Opalach, J.J. 1984. A Test of Consumer Demand Response to Water Pricing: A Reply. *Land Economics*, 60(4):416–419.

Osborn, C.T., J.E. Schefter, and L. Shabman. 1986. The Accuracy of Water Use Forecasts: Evaluation and Implications. *Water Resources Bulletin*, 22(1):101–109.

Polzin, P.E. 1984. The Specification of Price in Studies of Consumer Demand Under Block Price Scheduling: Additional Empirical Evidence. *Land Economics*, 60:301–309.

Porges, R. 1957. Factors Influencing Per Capita Water Consumption. *Water and Sewage Works*, pp. 199-204.

Rogers, Gladwin, and Harmony. 1986. *Garden Place Apartments: Water Conservation Plan*. Prepared for the City of Mesa, Ariz.

——. 1987. *Water Conservation for Commercial/Industrial Landscapes Through Xeriscape Conversion*. Report for Arizona Department of Water Resources, State of Arizona.

Schneider, M.L., and E.E. Whitlatch. 1991. User Specified Water Demand Elasticities. *Journal of Water Resources Planning and Management*, 117(1):52–73.

Southern Arizona Water Resources Association. 1993. *Landscaping to Save Water—Xeriscape, Tucson, Arizona.*

Stevens, T.H., G. Adams, and C. Willis. 1985. The Specification of Price in Studies of Consumer Demand Under Block Price Scheduling: A Comment. *Land Economics*, 61:327–329.

Taylor, L.D. 1975. The Demand for Electricity: A Survey. *Bell Journal of Economics*, 6(1):74–110.

Tucson Water. 1989. Unpublished water use data.

——. 1993. *Drip Irrigation: A Practical Approach to Landscape Watering in Tucson*. City of Tucson Water Conservation Office.

Williams, M. 1985. Estimating Urban Residential Water Demand Under Alternative Price Measures. *Journal of Urban Economics*, 18(2):213–225.

Witcomb, J.B. 1990. Water Use Reductions From Retrofitting Indoor Water Fixtures. *Water Resources Bulletin*, 26(6):921–926.

Young, R.A. 1973. Price Elasticity of Demand for Municipal Water: A Case Study of Tucson, Arizona. *Water Resources Research*, 9(4):1,068–1,072.

Index

NOTE: An *f.* following a page number refers to a figure; a *t.* refers to a table.

Alternative scenarios, 149, 156
American Statistical Association, 94
American Water Works Association
  appliance survey, 97
  Water Industry Data Base, 13–14
Autocorrelation, 31–32
Autoregressive integrated moving-average models. *See* Box–Jenkins models
Averages
  in causal/structural models, 56, 57*t.*
Averaging procedures, 74–78
  deseasonalizing water use, 78
  monthly series, 74*t.*, 76*f.*
  moving-average (13-month) method, 77–78, 77*t.*, 79*t.*
  technique, 74–77, 75*t.*

Backcasts, 67
Bias, 150–151. *See also* Unbiased estimates
Box–Jenkins models, 34, 87–88
Bureau of Economic Analysis, 93–94
Bureau of Labor Statistics, 94, 95
Business cycles, 147–148

Causal/structural models, 7, 53–67, 146–147, 148
  averages and standard deviations, 56, 57*t.*
  coefficient confidence intervals, 60–61
  correlation matrixes, 57, 58*t.*
  data, 55–56
  in estimating water conservation effects, 140–142
  estimating with time-series data, 63–65
  evaluating forecasts, 66–67
  explanatory variables, 55
  exploratory data analysis, 56–58
  frequency distribution, 57, 57*f.*
  improving accuracy with disaggregation, 65–66
  intercept dummy variable, 66
  regression analysis, 58–60, 59*t.*
  sample size, 54–55
  significance levels of the estimated coefficients, 60
  single-household forecast, 61–62
  slope dummy variable, 66
  total population forecast, 62–63
  using cross-section, time-series data, 66
Climate. *See* Weather and climate
  and water demand, 4–5
Clothes washers, 97
Coefficient models, 37–39
  adjusting, 48–49
  comparing per capita and per customer, 46–49, 47*t.*
  per capita, 37, 38*t.*, 39–44, 40*t.*
  per customer, 45
Coefficient of determination, 41
Coefficients
  confidence intervals, 60–61
  significance levels, 60
Cohort-component techniques
  in population forecasts, 92
Commercial and industrial use, 15–16, 16*t.*
  data, 19
  office space and floor area data, 20
Confidence intervals, 21–22, 29–30
Consumer price index, 93, 95
Control group, 137

Correlation coefficients, 102–103
Correlation matrixes, 57, 58*t.*
CPI. *See* Consumer price index
Cross-section data, 18
  pooled with time-series data, 18
Curve fitting, 25–35
  Box–Jenkins models, 34
  defined, 25
  exponential smoothing, 34
  goodness of fit, 26–27
  graphing, 25, 26*t.*
  linear regression, 25–26
  mean square error, 26, 27
  regression analysis, 27–34
  and seasonality, 34
  and time, 25
  and trend, 34
Customer billing records, 17

Data diskette, 9
Data on water use
  basic statistical analysis, 20–21
  cleaning, 20
  collecting, 17
  commercial and industrial data, 20
  cross-section, 18
  customer feedback, 155–156
  direct collection methods, 155–164
  disaggregation, 18–19
  employment data, 19
  income data, 19
  missing values, 20
  organizing, 17–19
  outliers, 20
  pooled cross-section and time-series, 18
  population data, 19
  surveys, 157–164
  time-series, 17–18
  weather data, 19
Degrees of freedom, 29
Delphi method, 156–157
Demand forecasting. *See* Water demand forecasting
Dependent variable, 28
Disaggregation
  improving accuracy in causal/structural models, 65–66
  principle of composition, 18
  principle of noise, 18–19, 54
  sectoral, 1
Dishwashers, 98

Economic cycles
  and water demand, 4, 147–148
Economic forecasts, 93–96
  accuracy issues, 94
  Bureau of Economic Analysis, 93–94
  checklist, 94–95
  income trends, 96
  and inflation, 95
  personal income, 95–96
Economic-base models
  in population forecasts, 92
Elasticity, 31, 49
Employment data, 19
End-use models, 6, 7*t.*
Error measures, 151
Evaporative coolers, 98–99
Evapotranspiration, 106–107
Expert opinion, 156–157
Exponential smoothing, 34, 78–83, 146
  example, 79–80
  Winter's exponential smoothing model, 80–83
Extrapolation, 7, 146
  in population forecasts, 92
Extrapolation models. *See* Exponential smoothing

F statistic, 30
Federal Home Administration, 117
Fees. *See* Rates
Forecasts
  evaluating, 66–67
Frequency distribution, 21
  in causal/structural models, 57, 57*f.*

Gallons per capita per day, 13
Garbage disposals, 98
Global warming, 109
Goodness of fit, 26–27, 41
gpcd. *See* Gallons per capita per day
Graphing, 25, 26*t.*

Heat-island effects, 108–109
Heteroscedasticity, 32

Income data, 19
Intercept dummy variable, 66
Irregular distributions, 22
IWR-MAIN, 8

Johns Hopkins Residential Water Use Project, 117

Land-use ratios
  in population forecasts, 92
Linear regression, 25–26
Lognormal distribution, 22

Mean, 20
Mean absolute deviation, 150
Mean absolute percent error, 150
Mean squared error, 26, 27, 149–150
Median, 20

Meters and metering, 115. *See also* Rates
effect on use, 117, 117*t.*, 118, 118*t.*
Microdata, 53-54
Modal value, 21
Mode, 21
Multicollinearity, 33
Multivariate weather and climate models, 104-106, 105*t.*

National Bureau of Economic Research, 94
National Weather Service, 101, 106
Nonlinear relationships, 33-34
Normal probability distribution, 28
Null hypothesis, 29

Omission of variables, 32-33

Parsimony, 33
Peak demand. *See* Seasonal and peak demand
Per capita consumption, 38*t.*, 40*t.*, 41*f.*
and customers served, 14, 15*t.*
by USEPA regions, 13, 14, 14*t.*
New York City boroughs, 15, 15*t.*
Per capita demand extrapolations, 6
Per capita models, 37, 38*t.*, 40*t.*, 147
compared with per customer models, 46-49, 47*t.*
exponential trend line, 44*f.*
forecasts, 44
formula, 41
linear trend line, 42*f.*
logarithmic trend line, 43*f.*
power-curve trend line, 43*f.*
trending use rates, 39-44
Per customer models, 37, 39*t.*, 40*t.*
compared with per capita models, 46-49, 47*t.*
forecasts, 45
value of, 46-47
Population
and water demand, 4
Population data, 19
Population forecasting
national, 89-91
national projection, 90*f.*
national trends, 91
and other demographic factors, 93
regional, 92-93
regional checklist, 92-93
regional trends, 91-92
Potential evapotranspiration, 106
Precipitation
graph, 103*f.*
Price elasticity, 115-117. *See also* Rates
commercial, 121
of demand, 115-116
in demand forecasts, 122-124
industrial, 120, 121*t.*
perfectly elastic, 116
qualitative findings of econometric studies, 119
quantitative findings of econometric studies, 120, 121*t.*
of revenue, 116
unitary elastic, 116
by US region, 120*t.*
Price
and water demand, 5
Price increase, 124
Price specification, 115
Principle of noise, 18-19, 54

R^2, 30
Rate schedules, 17
Rates, 111-112. *See also* Meters and metering, Price elasticity
constant commodity charge, 112, 112*f.*
decreasing block, 112-113, 113*t.*, 114*f.*
econometric studies, 118-121
fixed fee, 112, 112*f.*
flat, 113-115, 117*t.*
increasing block, 113, 113*t.*, 114*f.*
metered service, 115
monthly service fee, 112
unmetered service, 113-115
Regression analysis, 27-34
autocorrelation, 31-32
in causal/structural models, 58-60, 59*t.*
confidence interval, 29-30
degrees of freedom, 29
dependent variable, 28
elasticity, 31
F statistic, 30
heteroscedasticity, 32
multicollinearity, 33
nonlinear relationships, 33-34
normal probability distribution, 28
null hypothesis, 29
omission of variables, 32-33
parsimony, 33
R^2, 30
reliable parameter estimates, 31
serial correlation, 31-32
specification of the model, 28
specifying variables, 28
standard error, 29
standard error of the estimate, 31
statistical significance, 28-31
t statistics, 29
two-tailed test, 29
unbiased estimates, 31
Regression approaches, 83-86, 148
coefficients of seasonal regression, 85-86, 86*t.*
graph of seasonal regression, 84*f.*, 85

Reliable parameter estimates, 31
Reverse-osmosis water filters, 98

Seasonal and peak demand, 73–74
  averaging approaches, 74–78
  exponential smoothing, 78–83
  regression approaches, 83–86
Sectoral disaggregation, 1
Serial correlation, 31–32
Skewness, 22
Slope dummy variable, 66
Specification of the model, 28
Specifying variables, 28
Standard deviations, 21
  in causal/structural models, 56, 57*t.*
Standard error, 29
Standard error of the estimate, 31
Statistical analysis
  confidence intervals, 21–22
  frequency distribution, 21
  irregular distributions, 22
  lognormal distribution, 22
  mean, 20
  median, 20
  modal value, 21
  mode, 21
  skewness, 22
  standard deviation, 21
  variance, 21
Statistical significance, 28–31
Structural models
  in population forecasts, 92
Surveys
  designing, 157–158
  by mail, 162
  nonresponse bias, 159
  questionnaire development, 158–159
  random sampling, 160
  sample size, 160–162
  systematic sampling, 160
  target population, 159
  by telephone, 163–164

t statistics, 29
Technology
  and water demand, 4
Technology forecasts, 96–99
  appliances, 97–99
  clothes washers, 97
  dishwashers, 98
  evaporative coolers, 98–99
  garbage disposals, 98
  and industrial water use, 97
  reverse-osmosis water filters, 98
  water softening, 98

Temperature
  graphs, 102*f.*, 108*f.*
Time-series data, 17–18
  cross-section, 66
  estimating in causal/structural models, 63–65
  pooled with cross-section data, 18
Treatment group, 139
Two-tailed test, 29

Unbiased estimates, 31
US Bureau of Census, 89–90

Variance, 21

Water conservation programs
  aggregate analysis, 137
  competition (antagonism) between, 133
  control group comparisons, 137–139
  estimating effects of with disaggregated
    causal/structural models, 140–142
  estimating water-savings potential, 134, 135*t.*
  evaluating, 131–132
  goals, 134–135, 136*f.*
  net benefits, 131
  studying effects of, 137–142
  synergism, 133
  types of, 129–130, 131*t.*, 133
  variability of impact, 134
  and water demand, 5, 132–136
Water demand
  and economic cycles, 4
  and efficiency and conservation programs, 5
  influences on, 4–5
  and population, 4
  and price, 5
  and technology, 4
  and weather and climate, 4–5
Water demand forecasting
  accuracy improvement strategies, 151–152
  alternative scenarios, 149, 156
  applications, 3–4, 3*t.*
  approaches, 6–7
  books and monograms on, 1–2
  constrained, 48–49
  defined, 1
  error measures, 151
  evaluating for accuracy, 149–151
  examples (data diskette), 9
  factors in, 2
  implementing, 8–9
  information derived from, 1
  interval vs. point, 148
  long-term, 3, 3*t.*, 147
  as management tool, 2
  medium-term, 3*t.*, 4, 147–148

Monte Carlo simulations, 149
reasons for, 152–153
regression-generated confidence intervals, 148
and regulatory requirements, 2
sectoral disaggregation, 1
selecting a method, 8
short-term, 3*t.*, 4, 146–147
steps in process, 145
unconstrained, 48–49
water use per capita, 1
and weather and climate, 101–109
Water Industry Data Base, 13–14
regions, 13, 14, 14*t.*
Water softening, 98
Water use
collecting data, 17
commercial and industrial, 15–16, 16*t.*
Denver, 16, 16*t.*
interrelation with temperature and rainfall, 102–106
New York City, 15, 15*t.*
per capita, 13–15, 14*t.*, 15*t.*
WaterPlan, 8
Weather and climate, 101
climatic trends, 108–109
correlation coefficients, 102–103
cycles, 107–108, 147
evapotranspiration, 106–107
global warming, 109
heat-island effects, 108–109
individual weather variables, 106
multivariate models, 104–106, 105*t.*
Weather
and water demand, 4–5
Weather data, 19
Winter's exponential smoothing model, 80–83